A Man of
Misconceptions

A Man of Misconceptions

THE LIFE OF
AN ECCENTRIC IN AN
AGE OF CHANGE

John Glassie

RIVERHEAD BOOKS

a member of Penguin Group (USA) Inc.

New York

2012

RIVERHEAD BOOKS

Published by the Penguin Group

Penguin Group (USA) Inc., 375 Hudson Street, New York, New York 10014, USA · Penguin
Group (Canada), 90 Eglinton Avenue East, Suite 700, Toronto, Ontario M4P 2Y3, Canada
(a division of Pearson Penguin Canada Inc.) · Penguin Books Ltd, 80 Strand, London WC2R 0RL,
England · Penguin Ireland, 25 St Stephen's Green, Dublin 2, Ireland (a division of Penguin Books
Ltd) · Penguin Group (Australia), 250 Camberwell Road, Camberwell, Victoria 3124, Australia
(a division of Pearson Australia Group Pty Ltd) · Penguin Books India Pvt Ltd, 11 Community Centre,
Panchsheel Park, New Delhi–110 017, India · Penguin Group (NZ), 67 Apollo Drive, Rosedale,
North Shore 0632, New Zealand (a division of Pearson New Zealand Ltd) · Penguin Books
(South Africa) (Pty) Ltd, 24 Sturdee Avenue, Rosebank, Johannesburg 2196, South Africa

Penguin Books Ltd, Registered Offices: 80 Strand, London WC2R 0RL, England

Library of Congress Cataloging-in-Publication Data

Glassie, John.
A man of misconceptions : the life of an eccentric in an age of change / John Glassie.
p.   cm
Includes bibliographical references and index.
ISBN 978-1-59448-871-9
1. Kircher, Athanasius, 1602–1680.   2. Scientists—Germany—Biography.
3. Intellectuals—Germany—Biography.   4. Eccentrics and eccentricities—
Germany—Biography.   5. Germany—History—1618–1648—Biography.   6. Germany—
History—1648–1740—Biography.   7. Science—Germany—History—17th century.
8. Germany—Intellectual life—17th century.   I. Title.
CT1098.K46G53   2012   2012026023
943.041092—dc23
[B]

Printed in the United States of America
1   3   5   7   9   10   8   6   4   2

Book design by Chris Welch

For Natalie

CONTENTS

Let us not, in the pride of our superior knowledge, turn with contempt from the follies of our predecessors. . . . He is but a superficial thinker who would despise and refuse to hear of them merely because they are absurd.

—Charles MacKay,
Memoirs of Extraordinary Popular Delusions and the Madness of Crowds, 1852 edition

APOLOGETIC FORERUNNER TO

THIS KIRCHERIAN STUDY

Sometime in the early 1670s an old Jesuit priest named Athanasius Kircher began to write a remarkable account of his early
life. It told how, by virtue of divine intervention and his own
bright mind, he'd come out of nowhere (a small town in the forested
region of what is now central Germany) and survived stampeding
horses, a painful hernia, and the armies of an insane bishop, among
other things, to take his place as one of the intellectual celebrities of
the seventeenth century.

As a general rule Kircher never ruined a good story with facts.
Although well-known for more than thirty seat-cushion-size books
on almost as many subjects, and for an apparent knowledge of almost everything, he was also known for a tendency to embellish
on his own behalf. (The other frequent complaint against him was
more like the opposite: people said he was inclined to believe any
spectacular story he heard.) His uncharacteristically slim autobiography, written either in his bedchamber off a college courtyard
in Rome or at a mountainside retreat called Mentorella, certainly

included as many exaggerations and omissions as great escapes and miraculous recoveries, but the main story he told was true.

In many cases the truth is even more remarkable than Kircher was willing to let on. He doesn't come out and say, for example, that he first gained the attention of the learned elite in the early 1630s, at least in part, by claiming to own a clock that was powered by a sunflower seed and a mystery-solving manuscript written by an Arab rabbi. Unfortunately for Kircher, by the time he died in 1680, his stature was fading. After his death, for the most part, until a recent revival of interest in this baroque polymath, the custom was to either scoff or avoid discussing him altogether. "It is not the writer's intention to enter into the vast and terrifying subject of Athanasius Kircher," one art historian assured his readers when, in the middle of a 1972 paper on Jesuit architecture, he found himself in the vicinity.

"Vast" is an appropriate word because Kircher's curiosity and intellectual pursuits were almost unfathomably widespread. He was genuinely and insatiably curious about the world around him, and from his established place at the Collegio Romano, the flagship institution of his order in Rome, he threw himself into the study of everything from light to language to medicine to mathematics. In his museum at the Collegio he not only displayed antiquities, artifacts, and curiosities from around the world (amassed with the help of Jesuit missionaries), but also demonstrated his own magic lanterns, speaking statues, vomiting devices, and, as legend has it, a single "cat piano."

This was the kind of man who pursued his interest in geological matters by lowering himself down into the smoking crater of Mount Vesuvius. He spent decades trying to decipher the hieroglyphic texts of ancient Egypt because he believed, along with many others, that

Kircher's museum at the Collegio Romano

they contained mystical wisdom passed down from the time of Adam. He examined all aspects of music and acoustics, and experimented with an algorithmic approach to songwriting. He was among the first to publish a description of what could be seen through a microscope.

Kircher was so prolific and so ingenious that he might have been remembered as a kind of seventeenth-century Leonardo. The problem was that he got so many things wrong, and this is also where the "terrifying" part comes in. Many of Kircher's actual ideas today seem wildly off base, if not simply bizarre. Contrary to Kircher's thinking, for instance, there is nothing occult or divine about magnetism. There is no such thing as universal sperm. And there is no network of fires and oceans leading to the center of the Earth. It's fair to say that from the viewpoint of modern science Kircher has been something of a joke.

Of course, modern science didn't exist in 1602, when Kircher was born, but he lived right through the age in which it began. The story of the so-called scientific revolution, a term that was coined only in the twentieth century, is already a cliché, in which magic and superstition were subdued by rational minds and the experimental method. For people who lived then, it was a lot more complicated than that. But there does seem to be a consensus that what transpired during the seventeenth century, give or take a few decades, somehow explains how we became modern, how we became who we are.

When Kircher was born, to pick the easiest example, almost everyone assumed the Earth was at the center of the universe; at the time of his death almost every educated man willing to be honest with himself understood that it wasn't. (There wasn't much opportunity to become an educated woman, and most of Europe's forty million peasants were not aware of the debate.) At the very least, as the cultural critic Lawrence Weschler once put it, "Europe's mind was blown."

Athanasius Kircher—an apparently silly man, a somewhat untrustworthy priest, an egomaniac, and an author who inspired one

American historian to write in 1906 that "his works in number, bulk, and uselessness are not surpassed in the whole field of learning"—is perhaps not the most likely subject for a biography. Then again, he can just as easily be characterized as an extremely devout person, a champion of wonder, a man of awe-inspiring erudition and inventiveness, who, one way or another, helped advance the cause of humankind. His "useless" books were read in the royal courts of London and Paris and in the settlements of New Spain, later called Mexico. They were read, and often funded, by popes and Holy Roman Emperors. And they were read, if not always respected, by the smartest minds of the time. A secret Jesuit adherent of the Copernican system in the aftermath of the Galileo affair, a debunker of alchemy at the time Isaac Newton became obsessed with the practice, a collaborator with the artist Gianlorenzo Bernini on two of his most recognizable works, and an influence on Gottfried Leibniz's thinking about the binary system, Kircher, or rather the story of his life, might provide some insight into how we got here after all. One of the biggest characters of all time, he was also surprisingly representative of his own.

Kircher wanted the world to be magical, and yet to make sense, and he believed in his special ability to make sense of it. He also wanted to be famous (not to die), and he began writing his memoir in his seventies as part of a larger effort to shore up his legacy; he understood that his reputation was in decline. Around the same time he published another book, not under his own name but under the name of a student, titled *Apologetic Forerunner to Kircherian Studies*, in which he defended himself from his detractors and reaffirmed his belief in the magnetic healing power of something called the snake stone. He was also engaged in a dispute with an English gentleman over who could truly claim to have invented the megaphone. It wasn't possible for him to know how it would all turn out: the question wasn't *whether* he would be remembered, since he couldn't have imagined obscurity for himself, but how well.

PART ONE

1

Incapable of Resisting the Force

A ccording to the memoir of Athanasius Kircher, even the circumstances of his birth were auspicious. And in a sense they were, if you choose, as he did, to leave out the witch hunt.

Kircher's mother was "the daughter of an upright citizen," his father a learned man with "expertise in expounding complicated matters." They lived in a hilltop town called Geisa, part of the old principality of Fulda, in a valley of the gentle and green Rhön Mountains. (Fulda was also the name of the small city at the center of the principality; the trip there from Geisa took about three hours on foot.) For a long time before Kircher was born in 1602, his parents were caught up in the conflict that had disrupted northern Europe since 1517, when Martin Luther nailed his ninety-five theses to the door of the Castle Church in Wittenberg. Catholics and the new Lutherans felt the special kind of hatred for each other that comes from a split within the same religion, as did the Catholics and the Calvinists, the followers of John Calvin.

The effects of the Reformation were especially ugly in the Holy

Roman Empire of the German Nation—neither Holy, nor Roman, nor an Empire, as Voltaire would later say. It was more like an agglomeration, to use his word, of three hundred more or less autonomous entities loosely organized under the auspices of the Hapsburgs in Vienna, relatives of the king who controlled Spain, Portugal, the kingdom of Naples, the duchy of Milan, and a great deal of the new world. The Holy Roman Empire, such as it was, included feudal lands, secular territories, free cities, Catholic abbeys, and prince-bishoprics with many overlapping interests and internal faith-based animosities.

Kircher's father worked as a magistrate under the ruling Catholic prince of Fulda, a man named Balthasar von Dernbach. In 1576, as Kircher described it, the prince was "driven out by the persecution of heretics into exile." The "heretics" who threw him out of office and drove him all of three miles away were mainly Lutherans fed up with von Dernbach's effort to re-Catholicize the region, which included throwing Lutherans out of office and installing people like Kircher's father in their place. Although Kircher's father "favored Balthasar's most just cause and defended him with all his might against the attacking heretics," he too was "vexed by the persecution of the heretics," to say nothing of "the insolence of the heretics," and was forced to leave his post.

Von Dernbach and his lawyers spent twenty-six years building the case for his reinstatement, which finally occurred in 1602, near the feast day of Saint Athanasius, who himself had been forced into exile for staunchly defending orthodox Christianity against a powerful heretical sect. The birth of the ninth Kircher child on this feast day, so close to such an important occasion, meant that the child would get an important-sounding name: *Athanasius* comes from the Greek word for "immortal"; *Kircher* is a variation on the German for "church."

Soon after von Dernbach regained power, he began to cleanse the principality not only of heresy but of the influence of the devil. Inquiries turned up a woman in her late thirties named Merga Bien, who, among other suspicious things, such as being the wife of von Dernbach's political opponent, had recently become pregnant for the first time in her fourteen-year marriage. In jail she was locked in a dog kennel for some time and forced to confess that her pregnancy resulted from sexual relations with Satan. After more than twelve weeks of detention, she and her unborn baby were burned alive in Fulda's courthouse square.

Many people from the surrounding countryside were subsequently taken from their homes or fields to be put on trial for witchcraft in the name of the one true religion. Every couple of months, as many as thirteen women and girls at a time were burned alive, sometimes at the stake, sometimes all together on a huge pyre after having been stuck through with red-hot skewers. More than two hundred people were executed before von Dernbach died and his administration was finally dissolved, around Kircher's fourth birthday.

The hope is that young Kircher was not exposed to these scenes, to the screaming or to the reek in the air. But from a modern point of view, daily life in Geisa had a harshness and a reek about it regardless. Although Kircher's family probably lived in the kind of half-timbered house that now evokes Old World charm, they also lived within short range of their animals and their own waste. Unsanitary conditions were exacerbated by the fact that, as one historian has put it, "Westerners at this time looked on water with great suspicion," though, given the number of waterborne diseases, they were probably right to.

Even in the house of someone of relative status, such as Kircher's father, who at one point also served a few years as Geisa's mayor, the first floor was often used to keep chickens and pigs, to slaughter and butcher animals for meat, as well as to wash clothes and stock provisions. The walls of the house were made by filling its frame with a mix of clay and straw, and the roof was thatched. On the second floor there was probably a kitchen that was filled with smoke, because in these houses typically there were stoves but no chimneys, and a main room, a *Stube*, with hard benches, where family members would sit, eat, sew, mend, study, pray, hang and shelve household items, and perhaps sometimes sleep. Candles stank because they were made of molded tallow, rendered from beef or mutton suet. Any lamps also burned animal fat. In the cold rooms on the floor above the kitchen and the *Stube*, for two adults and nine children there were perhaps three straw or wool beds, though the numbers changed over time: two of Kircher's older brothers died in childhood.

The family's house was steps away from Geisa's market square, at the top of the long hill that led to the center of the town. The variety of available goods must have been fairly limited. People lived chiefly on soups, porridges, hard brown bread, and some meat, though Catholics ate fish and vegetables on fast days, particularly during Lent. When Kircher was a boy, potatoes, coffee, tea, tobacco, chocolate, and corn, not to mention the fork, were very likely unknown to him. (By the time he died, he had become an advocate of tea from China as treatment for kidney stones and hangovers.)

Despite the difficulties inherent in the way of life, the hilltop town of Geisa, overlooking the valley of the Ulster River, would have been a pleasant place to grow up. And from an early age, as might be expected of a boy who would later call himself "master of a hundred arts," Kircher displayed a "not ordinary aptitude" for learning. This was thought to go along with his somewhat earthy complexion, his dark skin and dark hair, coloring believed to indicate an excess of black bile, called melancholia. Melancholic types were said to be pensive, dreamy, and intellectual, suited to deep study and the attainment of knowledge, even capable of genius. Moreover, Kircher's head was large, generally agreed to suggest, as one seventeenth-century writer put it, "a wonderful intellect and a most tenacious memory." Kircher's father, a scholar of philosophy, theology, and rhetoric, who kept "thousands of books," apparently took an aging parent's interest in his youngest, most promising child. As his older sons "entered orders of various religions and daughters were joined in matrimony," old Kircher taught the boy music, Latin, and the fundamentals of geography—or as Kircher later described it, the study of "the world according to its divisions."

When not receiving lessons from his father, or from a rabbi his father hired to teach him Hebrew, he seems to have received attention from above. Young Kircher was athletic but accident-prone, a bit of an absentminded professor even as a boy, and sometimes he got into the kind of trouble that, he claimed, only the Virgin Mary could get him out of.

One hot summer day, he and some friends walked down to the bottom of the hill to cool off in the river. "It happened that in the midst of a certain mill house, the course of the river, in the manner of a lofty waterfall, was flowing with a more swift current because of the colossal trough of the mill wheel," he recalled. "Carried to this trough by boyish ignorance and snatched up with the current, I was completely incapable of resisting the force, and now closer and closer to the wheel, with the name of Jesus and the customary prayer to Mary, I trembled at the danger of death and the grinding of my entire body."

The friends who saw him being "snatched beneath the wheel head-long" all gave up on his survival, "especially since the wheel missed the bottom of the channel by so little that my body would scarcely be able to sustain itself without the pulverizing of all my limbs." When they finally found him downstream they were hardly able to believe their eyes: "By the singular protection of God and the Virgin Mary I emerged safe from the other side in such a way that no sign of harm was apparent on me." Having been "restored thus by divine mercy" to them, Kircher rejoiced with his friends and they all "acknowledged the apparent miracle."

Another apparent miracle occurred during an annual horse race one Pentecost Sunday. After a procession to bless the fields "against the storms brought in by screech owls and Satan," a crowd of people packed in tight to watch the event. "Upon the start of the race, individuals in the commotion were pressing one another more and more vehemently in their desire to see," Kircher remembered. "But I, merely a boy, stood in the front of the crowd, since I was not able to withstand the force of those pressing; and violently shaken from my spot into the stadium, I was rolled into the very whirlwind of the horses running with all their might."

The crowd all shouted for the horses to stop. But as Kircher later asked, "Who can stop galloping horses?" There was nothing to do but curl himself into a ball and entrust himself again to God and the Virgin Mary. "And since I had been lost in the cloud of stirred-up dust, everyone was sure that I had been ground to pieces by the stamping of the horses. But in reality, after the horses had run by I stood up unharmed and safe by the singular gift of God."

Many people gathered around him to wonder how "amidst so great a whirlwind" he had managed to preserve himself from danger. "To these I responded that not small was the power of the one who rescued Jonah from the belly of the whale, and Tobias from the devouring of the fish, and Daniel from the lions, and who kept me safe from the stamping of the horses."

There was something special about this boy, as everyone around and even the boy himself—or especially the boy himself—could see.

SOME YEARS before Kircher was enrolled in the Jesuit school at Fulda, at around age ten, a Protestant professor from Heidelberg complained about the relatively new religious order. "Very many who want to be counted as Christians send their children to the schools of the Jesuits," he explained. "This is a most dangerous thing, as the Jesuits are excellent and subtle philosophers, above everything intent on applying all their learning to the education of youth."

To Protestants, the growth of the Jesuits made for a threatening counterinsurgency, one all the more insidious for targeting the hearts and minds of young people. The Catholic order hadn't been created to combat the heresy of Protestantism but seemed particularly well suited to that task. Its founder, Ignatius of Loyola, a previously vain and self-absorbed mercenary from the Basque region of Spain, had discerned his calling after his leg was shattered by cannon fire at the battle of Pamplona in 1521. He employed military terminology when establishing what he called the regiment or the company of Jesus, whose members served as "soldiers of God" doing "battle with evil." (Later, more decorously, they were called the Society of Jesus.) Their official mission: "propagation of the faith"—"particularly the instruction of youth and ignorant persons in the Christian religion." And like young men signing up to fight an apparently righteous war, thousands of young Catholic men responded to the call to defeat the heretics, wielding a form of intellectual as well as spiritual vigor. In little more than sixty years since the first Jesuit school was started in the Sicilian city of Messina, more than five hundred schools and almost fifty seminaries had been established across Europe. There were about sixteen thousand members of the order by 1600 and Jesuit missionaries all over the world in such places as India, China, the Philippines, Congo, Ethiopia, Morocco, Brazil, Paraguay, and Canada.

All of this martial energy notwithstanding, the Jesuits exuded culture and sophistication. After his conversion, Ignatius had studied at the University of Paris, where he'd been exposed to the humanist values of the Renaissance. In the presence of natural beauty, he sometimes found himself in reverie. God existed in all things, according to Ignatius, and there was no reason to be cloistered away from the full realm of his creation. With a style that one modern historian has called "cosmopolitan, nonconformist, elitist," the Jesuits engaged young men in Roman and Greek classics, history, literature, and theater. They were so good at connecting with their students that one Protestant preacher believed they must "anoint their pupils with secret salves of the devil, by which they so attract

Fulda as it looked in the seventeenth century

the children to themselves that they can only with difficulty be separated from these wizards." And "therefore, the Jesuits ought not only to be expelled but to be burnt, otherwise they can never be gotten rid of."

Kircher's own favorite Jesuit teacher through several years at Fulda "concerned himself with this one thing, that to my passion for books I add a passion for piety." He also tried to prevent Kircher from being "misled by consort with depraved students." Boys like him were "privately summoned" by this priest, and encouraged "with all of his skills" to follow the lifestyle of the saints. "For these values then to such a degree did he inflame us in private conversations," Kircher recalled, "that for no other thing beyond the divine did we seem able to long."

Although Kircher later claimed to "spurn all those things that older boys are accustomed to do," he continued to be prone to misadventure as a teenager. "I had heard that a tragedy was being staged in a town two days' journey from Fulda," he remembered, "and, as I was curious and eager to see these sorts of things, together with my friends I entrusted myself to the journey." After the performance,

Kircher began the trip back by himself. The route led through part of the dense and damp Spessart Forest, which he described as "altogether horrible and infamous not only for its thieves but also for its host of dangerous wild beasts."

The danger was real enough in this and other wooded tracts of Europe. Robbers were said to kill their victims first and to check their pockets later. One man who traveled through German lands reported that, when caught, the criminals "are racked and tortured to make them confess, and afterwards their executions are very terrible." He saw many "gallows and wheels where thieves were hanged, some fresh and some half rotten, and the carcasses of murderers, broken limb after limb on the wheels." The infamy of the Spessart Forest in particular carries over into the twenty-first century, though today travelers through the region are more likely than anything to stop and enjoy a "Spessart Robber Buffet" along with a little Oktoberfest music.

"As soon as I had entered this forest, confused by the multitude of ways," Kircher wrote, in language that increasingly evokes spiritual searching, "behold, the more I progressed, the more I noticed that I was wandering from the true path, until, ensnared by brambles and thorn-bushes, I had no idea where in the world I was." The sun was going down, increasing his anxiety, and he began to lose hope of finding his way. His solution was first to entrust himself to God and to the Virgin Mary—and then to climb the highest tree. He spent the entire night in its branches, "safe from the wild animals" but nevertheless "in constant prayer." When morning came, after climbing down, he passed many hours in confusion and frustration. "Although I was not able to proceed because I was exhausted by the desperation of my soul and still more by hunger and by thirst, with new prayers addressed to God, I continued." Finally he came across a large meadow where some reapers were working; it turned out that after two days of wandering he was just as far away from home as when he'd started.

To Kircher, this incident—which ended with the reapers leading him through the region in exchange for an "exceedingly satisfactory recompense"—was yet another sign of God's "divine goodwill" toward him. It was not lost on him that, including the incidents with the mill wheel and the running horses, he had been saved a trinity of times. And so in return, he wrote, he "became wholly devoted to attaining a purpose in life and forfeiting things worldly."

It makes for an appropriate beginning to the official life of a devout Jesuit priest. To the extent that he believed he'd been the recipient of divine mercy, however, Kircher nurtured rather than forfeited the feeling that he was meant for something great.

BY THE TIME Kircher completed his secondary studies, at about the age of sixteen, he yearned to begin the two years of novitiate training, the three years of philosophy, the five years of teaching and practice, and the four final years of theology required to become a Jesuit priest. In fact, like his brothers, who had all joined religious orders, Kircher had limited freedom, economic or otherwise, to choose a different, or better, path. The Jesuits gave him an outlet for his religious zeal and his intellectual curiosity. Because they valued learning, they could accommodate what Kircher described as "a spirit unrelentingly devoted to acquiring knowledge" like his, though it was probably the prospect of traveling to some impossibly exotic and recently discovered part of the world as a missionary that ap-

pealed to him most. Convincing a Calvinist of the truth of transubstantiation was nothing compared with dying a martyr's death in a
place like Japan or New France.

Kircher was studying at the Jesuit college in the old city of Mainz,
where the Main River meets the Rhine, when he finally got news
of his acceptance as a candidate for the priesthood. But his "exceptional joy" didn't last long: "Barely had I received permission to enter
the Society from the Chief Provincial when, behold, most merciful
God wished to exercise his devoted servant with new tribulations."

In the winter of 1618, as he recalled, "all the rivers seemed frozen
with ice," and the broad Main provided a place for would-be Jesuits
with an inflated sense of self to put their ice-skating skills on display.
One day, Kircher remembered, with his usual lack of brevity, "I set
out with my friends in order to frequent the games which were accustomed to take place on the ice at that time of year, with the intention of showing my agility. And as I was rather desirous of glory, with
my agility and swiftness I was skating circles around the others,
burning by all means in boyish vanity to snatch the palm of victory.
It happened that after various displays of my skill I was striving to
surpass one of my friends who was more agile than I. But when, for
all of my exertion, I was not able to control myself, with legs splayed
and spread asunder, I struck the ice."

The result: a severe hernia.

And that wasn't his only health problem. "Added to this," he
wrote, "was a dangerous scabies of the legs which I had incurred at
nearly the same time from the chill and the sleepless nights that I
was spending at my studies." As a Jesuit historian has suggested,
these "scabies" don't sound like actual scabies, which are caused by
the burrowing of the itch mite under the skin to lay her eggs. More

likely, Kircher had a case of chilblains: ulcers of the skin from expo-sure to the cold. Either way, he worried that permission to enter the order, the object of so many "ardent pleas," might be withdrawn if these medical conditions were found out. It was well known that the Jesuits weren't just looking for pious, well-spoken, naturally talented young men; they were looking for physically robust candidates with-out, according to Ignatius, "stomach trouble or headache trouble or trouble from some other bodily malfunction." A "lack of bodily in-tegrity, illness or weakness" could disqualify a candidate, as could "notable ugliness," since it did "not help towards the edification of neighbors."

"And thus," Kircher explained, "lest the diseases become known to my superiors, I resolved to conceal each with utmost silence." He lived with these maladies for some months without getting treat-ment or discussing them with anyone—except God, "to whom alone my sufferings were known."

Over the course of these months, Kircher must have heard the news from Bohemia: Protestants had thrown two representatives of the Catholic Hapsburg emperor out a window of Prague's Hradčany Castle. A pile of garbage broke their fall. Since this "defenestration of Prague," as it became known, the Bohemian estates had organized in official rebellion against their own king, Ferdinand, who was next in line to become Holy Roman Emperor. By the time Kircher was supposed to begin his novitiate at the Jesuit college of Paderborn in Westphalia, preparations were being made for the Thirty Years War, though it would be thirty long years before anyone would think to call it that. "Meanwhile," Kircher recalled, "with every passing day the hernia was growing, and the scabies were worsening at a spec-tacular rate."

Paderborn sat 170 miles north of Mainz, on the other side of a mountainous region called the Hochsauerland. It would take more than a week to cover that distance under the best of circumstances, and a trip through the German provinces during the early seventeenth century could be challenging, to say the least. One Englishman and his brother traveling from Hamburg to Prague were "carried day and night in waggons" to a town called Hildesheim, then walked 130 miles or so to Leipzig, where the coachmen were too afraid to travel into Bohemia because of the war, so they hired "a fellow with a wheelebarrow" to carry their things ("our cloakes, swords, guns, pistolls, and other apparell and luggage") on a two-day trip to Penig, from which they traveled by cart, then wagon, then foot again to reach their destination. Otherwise, as this traveler facetiously described it, the journey was all "excellent cookery" and "sweete lodging." They usually slept "well littered" in the straw of a stable, and when taverns didn't serve "pickl'd herring broth" or "dirty pudding" or a raw cabbage "with the fat of rusty bacon poured upon it," they might offer "Gudgeons, newly taken perhaps, yet as salty as if they had beene three yeares pickled, or twice at the East Indies, boyled with scales, guts and all, and buried in Ginger like sawdust."

Whatever the details of Kircher's own trip—"Only He who knows the hearts of all knows how many difficulties I, afflicted by so much suffering, endured on that journey"—it worsened the problem with his legs. After finally arriving at Paderborn, he remembered, "since I tottered in my gait from the enormous pain in my feet, I was compelled to uncover my disease to my superiors, who noticed it." A surgeon was called to examine his legs. Gangrene had set in, and the doctor "immediately declared it incurable."

Still concealing his hernia "with the utmost silence, lest there be a fuss over me when two incurable diseases had been discovered," Kircher was informed that he would be dismissed from the Society if he wasn't better in a month. "Nothing remained except the Virgin Mary, the sole refuge of my health," he wrote. "And thus, in the dead of night at the foot of an exposed statue of this very Virgin, I lay prostrate on the ground with tears, as though She were the sole curatrix of the human race; and by such means and passions that I leave to the conjecture of the reader, I, her most downcast son, entreated the Great Mother with vehement prayers."

When Kircher woke up the next morning he discovered that his legs were "completely healed." And that wasn't all: "I also noticed that my hernia had vanished." The surgeon "proclaimed it a miracle," and Kircher's superiors "praised God and the Holy Mother Herself, by whose beneficence and intercession so marvelous a cure had occurred."

There is no satisfaction for the skeptical in Kircher's autobiography. "By confessing these things in this place," he wrote, he intended only to "spread about and stir up the honor and worship of God and the Blessed Virgin within my fellow men."

The Virgin Mary may have taken a more laissez-faire position on other matters, like this teenager's subsequent transition to daily life at the seminary in Paderborn.

2

Inevitable Obstacles

Kircher's real initiation into the Society of Jesus began in the middle of one night that autumn, when he was stirred from sleep and the warmth of whatever coarse blanket the seventeenth-century Jesuits of Paderborn could provide. In candle or lamplight, a priest explained the meditation Kircher was about to make, the first of many over the next four weeks. All novices were (and still are) led through the Spiritual Exercises of Ignatius of Loyola, but he wouldn't have been told exactly what to expect. Refined by Ignatius over years of rigorous contemplation, the exercises were meant to clear away the complicated layers of self-interest that normally motivate life choices, as well as the interests of evil, in order to discern the interests of God. For these young religious soldiers, it functioned like a spiritual boot camp; the process involved tearing down mental habits and assumptions and building up a desire to "freely choose" devotion and obedience in their place.

The first meditation began at midnight, according to Ignatius's written directions, when novices were to consider the "gravity and malice" of sin and to ask God for the "personal shame and confusion" appropriate to sinners. Through the day they cataloged and contemplated the "intrinsic foulness" of each and every sin they had ever committed, and an hour before supper they were instructed to imagine hell with all the senses: to see "the great fires" and the "souls appearing to be in burning bodies"; to hear "with one's ears the wailings, cries, howls, blasphemies"; to "smell the smoke, the burning sulphur, the cesspit and the rotting matter"; to taste "bitter things, such as tears, sadness and the pangs of conscience"; to feel "how those in hell are licked around and burned by the fires."

Novices repeated the meditations, prayers, and colloquies of the first day each day for the first week. The time was spent in silence and solitude, except for recitations and talks with the priest giving the exercises. Eyes were covered and the doors and shutters were closed to keep out the light. Having conjured such graphic scenes of the hell that awaited them, novices often wanted to do penance for their sins—depriving themselves of heat, food, or sleep, and "chastising" their bodies by wearing haircloths or striking themselves with cords or chains. "The most practical and safest in regard to penance seems to be that the pain should be felt in the flesh and not penetrate to the bone," Ignatius advised. "Therefore, the most appropriate seems to be to strike oneself with thin cords."

In the second week, novices spent hours a day applying their minds and their senses to the story of Christ's life and good works. In the third week, to the blood and tears of his passion. And in the last week, finally, with joy and gratitude, unshuttered windows and sunshine, to his resurrection. As meditations progressed, novices were urged to compare their own previous choices and actions with those of Jesus, and to contemplate the kind of life that God intended for them—they were to try to sort away all other voices, influences, and impulses. Discerning God's call from the deceptive call "practiced by the evil leader," otherwise known as the devil, was difficult, especially since the evil leader often tempted people "under the appearance of good."

The key was to get rid of what Ignatius called "disordered attachments" to such things as comfort, success, and praise. Losing these attachments required a special kind of humility: "I have it," he explained, "if I find myself at a point where I do not desire, nor even prefer, to be rich rather than poor, to seek fame rather than disgrace, to desire a long rather than a short life, provided it is the same for the service of God and the good of my soul." And yet the most perfect humility is achieved when you actually "want and choose poverty with Christ rather than wealth, and ignominy with Christ in great ignominy rather than fame," and when you "desire more to be thought a fool and an idiot for Christ, who was first taken to be such, rather than to be thought wise and prudent in this world."

It can't be said whether Kircher, still a teenager, actually discerned God's call, or whether he believed that he had, or whether these intensive days of meditation and fasting brought on any of the psychological or physical symptoms—euphoria, light-headedness—that might be mistaken for the effects of spiritual revelation. But these exercises had a profound influence on him. In order to achieve salvation in heaven—his ultimate desire, and a preference or self-interest that no one, not even Ignatius, had chosen to give up—he understood that he was going to have to develop some humility here on earth. And soon, it seems, he took pride in showing more humility than any of his peers. Kircher devoted himself to the ecclesiastical and communal life, and after making his first vows of poverty, chastity, and obedience at the end of the two-year novitiate, he began the three-year program in philosophy. "I did not dare to reveal my talent of intellect," he wrote in his memoir, "fearing lest, from the complacency arising from some degree of vainglory, I would diminish the flow of divine gifts into me." The decision had the added benefit of bringing on the kind of scorn that, as described by Saint Ignatius, made him seem even *more* like Christ. "This silence and masking of my ability caused both the instructors and the students to consider me stupid," Kircher remembered. In fact, they "all judged me to be foolish and stupid in my rejoicing in and exultation of the love of Christ."

In Kircher's retelling, he not only took his spiritual life more seriously than the other novices did, but actually had something significant—"the great gift of my intellect"—to be humble about. But if one of the most verbal men in history really kept silent, as he claimed, throughout his course in logic, an entire year of disputation and oral argument, it was in fact a profound act of self-restraint. And if he continued "in this manner" the next year, when studies turned to "physics," or natural philosophy, now called physical or natural science, he must nevertheless have been paying attention; he devoted many of the next sixty years to it.

FOR THE HUMANIST JESUITS, there was no conflict between religion and knowledge of the natural world. A greater understanding of the physical cosmos made for a greater appreciation of God's beautiful, complex creation, and a greater love for God, especially since, as the long-held belief went, everything in the earthly realm was connected through a great chain of being—through graduated correspondences and affinities—to the celestial realm above.

The Catholic authority on theology was Thomas Aquinas, for whom the authority on "physics" was Aristotle, for whom the universe was perfect and finite. All things not only had substance and form but a "sake," or final cause, or nature. The final cause, or nature, of an acorn, as the well-known example went, was to be an oak tree. It was the nature of things that were composed of earth, such as stones, to fall down toward the center of the earth. Water sought its place at the earth's surface, air sought its place above the earth, and fire sought its place above the air. Anything that didn't behave according to its form, or substance, or its presumed nature, was given either a different underlying nature or a hidden virtue.

Earth itself was fixed at the center of the cosmos, and beyond the realm of the earth everything was made of a fifth element, or "quintessence," called ether. The perfect spheres of the sun, moon, and the *five* planets, as well as the fixed stars, were contained within their own perfect celestial spheres, which, with the help of certain "intelligences," were moved around Earth in perfect circles. The final cause of these spheres: to be moved by the divine intelligences as objects of love.

According to Aristotle, nature abhorred a vacuum. The speed at which a thing fell was inversely proportional to the density of the medium it fell through—the lesser the density, the faster the thing fell—so a void or vacuum could not exist without having everything fall through it at infinite speed. Infinity itself was also not something that could exist, in part because anything infinite would have to be composed of things that were themselves finite, sensible, and therefore measurable.

As the old story of the dawn of the modern age goes, Aristotle was the figure who had to be toppled by the new men of science and reason. But it wasn't that established Aristotelian ideas sounded strange; nothing felt more intuitively right, for example, than the idea that the ground you stood on was immobile and at the center of things, and that it was the sun that moved across the sky. Although published by Copernicus in 1543, *De Revolutionibus* (*On the Revolutions*) had been condemned only recently by the Church.

The Jesuit professor of philosophy "shall not depart from Aristotle in matters of importance," instructed the order's plan of studies. "He shall be very careful in what he reads or quotes in class from commentators on Aristotle who are objectionable from the standpoint of faith." But adhering to Aristotle wasn't as straightforward as it sounded. Over the centuries, hundreds of commentaries on Aristotle's dozens of works had been produced. The eight-volume effort by Jesuits of the university at Coimbra in Portugal, published in many official and unofficial and fraudulent editions, was used to consider philosophical complexities related to, say, astrological influence on Earth.

Although the use of astrological study for divinatory purposes (forecasting) had been condemned by the pope in 1586, it was otherwise an integral feature of astronomical study. Almost no one imagined a world in which some kind of astral influence wasn't exerting itself. If not, what were the planets and the stars for? And why bother banning a futile endeavor? Any number of assumptions went unchallenged. No one thought, for example, to doubt the concept of spontaneous generation; it was simply assumed that small creatures such as worms, flies, ants, and even frogs and snakes, grew from nonliving matter, particularly if swampy or putrescent or excremental. Who could deny that maggots appeared on rotting flesh? "It be a matter of daily observation," as one seventeenth-century writer characterized it, "that infinite numbers of worms are produced in dead bodies and decayed plants." And plenty of the old authorities, including Aristotle, agreed. (Aristotle believed in spontaneous generation from *living* matter too, contending that cabbages engendered caterpillars.) Augustine surmised that *semina occulta* (hidden seeds) were responsible for things that sprang up from the earth "without any union of parents." Pliny held that insects originated from rotten food and milk and flesh, as well as from fruit, dew, and rain. Ovid, Plutarch, Virgil, and Democritus all apparently believed that bees were born from the dung of bulls.

Aristotle didn't stand in the way of modern science all by himself. And it wasn't impossible or uncommon, especially outside the Church, to disagree with Aristotelian ideas. But natural philosophy was philosophy: it was more through reason than observation that the natural world was known; through erudition, as opposed to experimentation, that the "truth" of a matter was usually determined. Ancient authorities held sway; the more ancient they were, the more sway they held. Knowledge increased by adding authorities, arguments, and commentaries onto the pile, rather than by ruling out ideas through trial.

KIRCHER DIDN'T GET very far in physics in Paderborn. Barely two months had been spent in the course when, as he put it, "a new crisis arose which presented to me the ultimate occasion to endure suffering and grief on behalf of Christ."

The crisis came in the form of Prince Christian of Brunswick, also known as Christian the Younger, the Insane Bishop, the Mad Bishop, and the Mad Halberstadter. This Protestant military leader sometimes referred to himself as *"Gottes Freund, der Pfaffen Feind"* (God's friend, the priests' foe), other times simply as "the supreme hater of Jesuits." Rallying to the increasingly complicated cause against the Catholic Hapsburg emperor, Christian had levied an army of ten thousand and was advancing through Westphalia, in the direction of Paderborn.

Christian had been made bishop of Halberstadt just a few years before, at the age of seventeen, after the death of his older bishop brother. "He possessed little qualification for this office," wrote twentieth-century historian C. V. Wedgwood, "save an unreasonable dislike of the Catholics." Christian wore a sparse mustache and an early modern mullet, and he was preceded in Paderborn by his well-cultivated reputation for committing unspeakable atrocities. "The most famous of them, namely that he forced the nuns of a plundered convent to wait, naked, on him and his officers, was invented by a journalist in Cologne." Nevertheless, he had torn through Westphalia in his own particular way. "He issued startling letters, suggestively burnt at the four corners, and bearing the words 'Fire! Fire! Blood! Blood!' to every sizable village he passed. This method seldom failed to extract a ransom in hard cash from the people."

As Christian's army approached, the Jesuit superiors acted to close the college—"lest there be a violent attack on the city and all be cut down to a man," Kircher explained. Soon a crowd of Paderborn's Protestants formed outside its doors; Christian's proximity apparently freed them to manifest their own hatred toward the Jesuits. When the rector went out to speak to the mob, a burning torch was thrown at him. He was beaten and dragged away. Inside the school, a plan was made for the priests and novices to leave that night in small groups. They were to change out of their robes and into secular clothes. "And since the enemy was beginning to encircle the city little by little," recalled Kircher, "and since the orderly was not able to offer provisions necessary for a journey, given the very sudden state of confusion," the priests and novices were "sent away whither Divine providence and fate might lead them. I together with three of my friends was among these."

Kircher and his companions made it out through the town gates. Avoiding the roads and making their way slowly, they hoped to reach the small city of Münster, home to another not always well-appreciated Jesuit community. It was located about fifty miles west, through thickly wooded lowlands. "The winter at that time was harsh and the snow was very deep, and, what was worst of all, we were poorly clothed and were lacking the necessary provisions," he remembered. "But the driving fear of the pursuing soldiers furnished wings to us in our flight."

A dozen or more of Kircher's fellow Jesuits were in fact captured. As a Catholic official later described it, Christian "took subjects prisoner, bound them, beat them, martyred some of them to death" and "similarly maltreated others." He seized all of the town's supplies, livestock, and grain, as well as its "cannons, munitions and silver plate." Christian himself later boasted that during the rape and pillage of Paderborn he'd probably fathered enough "young Dukes of Brunswick" to keep the priests in line there for a generation to come.

Kircher's group "wandered in the most dense forest and fields" all through the night and into the next day. "Though immersed up to the knee in thick clods of snow, we were progressing as much as we were able on our journey through this harshest wilderness," he recalled. He was so hungry, he claimed, that he gladly would have experienced "the degree of pleasure afforded by roots and grass, were the depth of the snow and the earth packed with ice not begrudging us this joy!" Finally, "frozen head to toe with trembling cheeks and faces turned completely blue," they found their way out of the forest and came across a cottage, where they were given some crude bread to eat. "It was of the worst type . . . made from straw and bran," he remembered. "Nonetheless it was as sweet to my famished palate as nothing that I can recall eating during my entire life."

Sometime that evening, in a place with poor dwellings and a fire burning in the dark, Kircher and his friends met up with "a certain man" who gave them warm food and a bed for the night. By the end of the next day they passed through the gates of Münster, which had seen its share of religious conflict. (Iron cages that were used decades before to display the corpses of executed Anabaptists still hung from the steeple of St. Lambert's church, and still do.) They recuperated there for about a week, until they heard Christian was moving in their direction, and then set off again, heading west (farther and farther from Paderborn, Fulda, Geisa, and home), toward the Rhine, about sixty miles away. There was yet another Jesuit college in the town of Neuss, not far on the opposite side. From Neuss, the more permanent safety of Cologne could be reached in about a day of travel south along the river.

After two more days of hiking, they came to the Rhine's east bank near Düsseldorf. The river, which runs all the way from the Alps to the North Sea, appeared to be frozen over. As Kircher later learned, the locals were usually willing to pay someone to find out if the ice was "solid enough to bear the weight of men and livestock." But with one look at Kircher's group, they recognized an opportunity to save some money: "Since they saw that we were poorly clothed (for we were dressed in secular garb), and since they strongly sensed that we desired to cross the river that very day, and since they speculated that we were men of little value, or fugitive soldiers, they believed that it would be of little consequence if they persuaded us, though we might die from it, to test the way."

Whoever these people were, they "happily and with the utmost mendacity" took the young men to the best place from which to cross. Kircher went first, treading carefully. His companions trailed behind in single file, some paces apart. "I then, as the leader of all, tested the way," he recounted, but "when now I reached the middle of the river, behold, I saw the entire Rhine exposed before me." His frightened companions began making their way back to shore, but he had "progressed farther than the solidness of the ice was bearing." Trying to follow his friends to the riverbank, Kircher saw that the ice was breaking up where they stepped, leaving him "in the middle of an island, as it were." Once his friends reached the bank, they dropped down on their knees to pray for his safety. As they prayed, they watched him being carried down the river, alone on his floating island of ice.

Kircher fled to God with his tears. As for his heart, "it held faith in God" and "it even knew that God . . . would never fail his own." Finally his ice floe was caught against "enormous heaps" of others building up into a jam of unsteady masses. "It was as dangerous as it was difficult to climb over this huge pile of ice fragments," he recalled. "Still, unless I preferred to die, it had to be attempted."

As it happens, most people now are familiar with an image of an icy Rhine near this spot: the nineteenth-century painting *Washington Crossing the Delaware*, by Emmanuel Leutze, who used the Rhine, not the Delaware River, as the basis for the scene. To judge from the way Kircher told the rest of the story in his memoir, however, his crossing was perhaps even more heroic than Washington's:

> Two altogether inevitable obstacles to passing over this heap presented themselves, the first of which was the slipperiness of the ice, which offered aid in climbing to neither feet nor hands. The other was the cracks, which had come into the fragments straight through to the surface of the water and into which, should I fall, there would be no human hope of escaping.
>
> What spirit I would possess in the face of so many unavoidable dangers, God alone knows. With fear nevertheless adding diligence to my nature, I went by the manifest aid of God through the smaller bits to the other part of the Rhine, where the river was drawn together by the more solid ice. While I continued thus on my way straight to a point somewhere near to the far shore, behold, I see the Rhine utterly opened. What I should do I was barely able to consider. Retreat was impossible, progress forward was difficult; however, that I remain there, exposed as I was to the harsh cold in the deep winter, completely exhausted by my sufferings, fear and anxiety of spirit, and, furthermore, wounded on my hands and feet by the sharp bits of ice, was nothing more than awaiting death itself. So, no other recourse remained for me but that I reach the opposite shore, which stood only about twenty-four feet away, by swimming (for as a boy I had learned to swim). The undertaking transpired thus: since amidst my swimming I was weighed down by my clothes, I tested for the bottom, and when I had found it, emerging now to my neck, a little after to my breast,

and now finally to my knees, I covered with ease the remaining distance.

Once on the far shore, although his limbs were "stiffened by the vehemence of the cold," Kircher fell to his knees and thanked God "for so clear a manifestation of divine protection." Slowly he began the three-hour journey to Neuss by himself. "With the assistance of divine grace I finally reached that town," he remembered, "where in the college I had already been announced by my comrades, who had crossed another part of the river, as dead and drowned."

3

A Source of Great Fear

Kircher was "received and restored" in Neuss "to the tremendous joy of all." After three days of rest, and another several hours of hiking on rutted, frozen roads, the young men finally walked wide-eyed through the gates of Cologne, a center of trade along the Rhine that had once rivaled Paris in size, sophistication, culture, and learning. A sovereign and heavily fortified "free city" within the Holy Roman Empire, Cologne laid claim to forty thousand citizens, an entire army of its own, one hundred fifty churches, and the world's largest incomplete cathedral: building had begun almost four hundred years before Kircher's arrival, but no work had been done for almost a century. Construction wouldn't resume for another two hundred years. A couple of streets away from the site of this unfinished Gothic mountain, as many as fifteen hundred students took classes at the Jesuit college.

Within this more urban setting, Kircher went on with his course in philosophy. Now he was the country boy in worn-out shoes, known or whispered to have barely escaped martyrdom at the hands of the Insane Bishop. He was still extremely pious, and still pretending to be a dimwit, but he wouldn't be able to pass himself off that way much longer.

Kircher and other Jesuit scholastics read Aristotle's works in Greek and discussed them in Latin. They also took general instruction in mathematics, a strange part, on the face of it, of a philosophy program meant to prepare them for theology and the priesthood—especially since mathematics had traditionally been viewed with condescension by natural philosophers and theologians alike. Mathematics could be used to measure and describe, and it had many practical applications, but it couldn't *explain*, in the opinion of natural philosophers, the way natural philosophy could. It had nothing to say about the causes or the natures or the essences of things, only about, as one historian put it, the "superficial quantitative properties" of things, properties regarded by philosophers as incidental. Mathematicians were bean counters, and their instruments (astrolabes, quadrants, protractors, plumb levels, calipers, magnetic compasses) just made for better bean counting.

But the reputation of mathematics as a field of study had improved a great deal in the sixteenth century. New translations of works by Archimedes of Syracuse on floating bodies and mechanics gave engineering its own seemingly infallible ancient authority. Mathematics included four traditional academic subjects: arithmetic, geometry, astronomy, and music (basically the study of harmonics). But it also encompassed mechanics, engineering, optics, surveyorship, and astrology. People in positions of power cared more and more about math as applied to architecture and construction, shipbuilding and navigation, mapmaking, fortification, armament, ballistics, and so on. And one Jesuit in particular had exerted himself to elevate the role of mathematics within the order. He was born Christoph Klau, in Bamberg, but his Latinized name was Christopher Clavius, and many of his contemporaries referred to him as the Euclid of his time.

Based at the prestigious Collegio Romano, the Jesuit college in Rome, from early in his priesthood until he died in 1612 (when Kircher was about ten), Clavius published works on astronomy, geometry, the construction of sundials, and algebra, a relatively new field in the West. He was among the first to use a dot or point as a decimal separator, parentheses to enclose calculations within an expression, and an *x* for variables. Most memorably, he recalculated the calendar year at the request of Pope Gregory XIII. The Julian calendar, put into use by Julius Caesar more than fifteen hundred years before, had been slipping to the point where it seemed that Easter would soon be celebrated in February. In order to start fresh with the new calendar, ten days had to be skipped outright. People in Catholic lands who went to bed on October 4, 1582, woke up the next morning on October 15. (Protestant countries resisted the switch for more than a century. Japan and Korea started using the calendar in the nineteenth century, and it took the Bolsheviks to make the change in Russia, which they did after coming to power in 1917. The Gregorian calendar is good until the year 4317, when a single extra day will have to be added.)

After suffering attacks by dubious Protestants who refused to adopt the new calendar—was this some kind of Catholic trick to steal time?—Clavius presented a plan to the Jesuit hierarchy by which the general intellectual reputation of those heretics could be, as he wrote, "most rapidly and easily destroyed." He urged the Society to identify and nurture those men who might become "outstandingly erudite" in what had previously been deemed the "minor studies of mathematics, rhetoric, and language." Clavius envisioned an elite corps of mathematician priests "distributed in various nations and kingdoms like sparkling gems," serving as "a source of great fear to all enemies" and as "an incredible incitement to make young people flock to us from all the parts of the world." Many of his proposals were put in place. And so while he was rigorously and rather inflexibly educated in Aristotelian and Thomist doctrines, Kircher also received private instruction in the very discipline that was beginning to undermine them.

In 1610, twelve years before Kircher arrived in Cologne, Galileo Galilei, a mathematics professor in Padua, published a slim volume, *Starry Messenger*, about the observations he'd made with a new instrument he called the *perspicillum*, or the telescope. It was a very much improved-upon version of the spyglasses that had recently begun to appear in Europe. The configuration (a concave lens at one end of a tube, and a convex lens at the other) was fairly simple, but Galileo's handcrafted device made things appear, as he wrote, "nearly one thousand times larger and over thirty times closer" than they would with the naked eye. Among other discoveries, he observed four moons revolving around Jupiter. The most basic implication of this was clear to any student of natural philosophy willing to admit it. (At the Jesuit school of La Flèche in Anjou, a student named René Descartes is said to have written a sonnet celebrating the news.) If moons revolved around Jupiter, maybe Earth wasn't really the center of the universe, around which everything revolved. Galileo also reported that "the moon is not robed in a smooth and polished surface," as Aristotelian doctrine had it, "but is in fact rough and uneven,

covered everywhere, just like the Earth's surface, with huge promi-
nences, deep valleys, and chasms."

The Church seemed willing at first to consider these findings.
Clavius and other Jesuit astronomers held a reception for Galileo in
Rome, and while Clavius declared telescopes "troublesome to operate," he confirmed the existence of moons around Jupiter. On the
question of the rough surface of the moon, however, the seventy-one-
year-old astronomer couldn't bring himself to believe his eyes. Perhaps the moon was just unevenly dense, he suggested. In a letter to
the Church's chief theologian, a group of Jesuit astronomers wrote
together that they were "not sufficiently certain" about the matter. In
other words, preconceived notions were such that Clavius couldn't
see through a telescope what modern people, who know the truth,
can recognize with the naked eye.

When Galileo was offered a new job in the Medici court in Florence around this time, he insisted on a double title: "mathematician
and philosopher." Since then he'd gotten into a dispute with Christopher Scheiner, another Jesuit in Rome, over who had been the
first to observe spots on the sun, and what they were. Galileo argued
that *he* had, that they were imperfections in another supposedly perfect sphere, and that their apparent movement was due to *Earth's*
orbit around the sun. In 1616, after certain Dominicans added their
own complaints against him, Galileo was admonished by the Church
for his Copernican views. By the time Kircher arrived in Cologne,
Galileo was in yet another public dispute with yet another Jesuit
astronomer, over the nature of comets.

During this same period, Johannes Kepler, a brilliant astronomer
born near Stuttgart, had been arguing his own mystical as well as
mathematical case for a sun-centered cosmology. Kepler spent years

working with reams of astronomical data he'd inherited, some say purloined, from the estate of astronomer Tycho Brahe, his former employer. He believed the planets were drawn around a living sun, in ellipses, not circles, by a spiritual, magnetic force. (Although a devout Lutheran, Kepler was currently in the service of the new Catholic Hapsburg emperor, perhaps not actually doing astrological predictions for his military leaders but supplying the astronomical readings for the astrologers who were.)

In this kind of environment, the Jesuits needed all the new mathematicians they could get. And despite Kircher's efforts to conceal his intellect, his professors in Cologne saw that he had a bent for mathematics. Humility was important, and generally it *was* better to want to live with Christ in ignominy rather than fame, to be thought a fool and an idiot with Christ rather than to be thought wise and prudent without him. But nothing was more important than to discern God's call properly. Some were meant to achieve great things, *ad majorem Dei gloriam*, as the Jesuit motto went, for the greater glory of God.

A YEAR AND A HALF LATER, twenty-two-year-old Kircher left Cologne and traveled down the Rhine to the small city of Koblenz. Going south along this route, the riverbanks grow higher and higher until they become tall promontories on top of which sit castles and ancient fortifications. Koblenz itself, whose name comes from the Latin for "confluence," takes up a triangle of land at the point where the Rhine and the Mosel rivers meet. Yet another Jesuit college and a brand-new church were situated in the middle of the town, helping to form a public square.

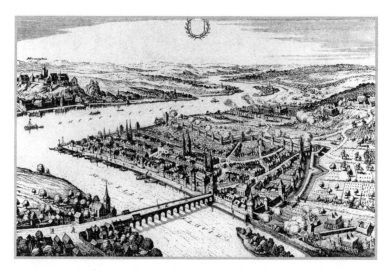

Koblenz, where the Rhine and Mosel rivers meet

It seems that while in the midst of his philosophy studies (in its third year, the program turned to metaphysics) and his special mathematics training in Cologne, Kircher had also shown enough proficiency with languages to be given a professorship in Greek. "The time came," he explained in his memoir, "when I was compelled out of obligation to reveal those talents that up until then had remained hidden." And this, he didn't seem to mind adding, was greeted with extreme surprise and envy: "People were unable to grasp how a man, for whom up to this point they had held no regard, and in whom no vestige of any inborn intellectual skill had been manifest, was able to evince those points which barely fell within the ken of the greatest masters of languages, mathematics and recondite wisdom."

People of Koblenz may have had trouble grasping other things, like what Kircher was doing up on a platform on the side of the

Society's church. They would have seen soon enough that this student of the works of Clavius was installing a sundial he'd designed for the location, complete with a Latin inscription that read, "See! How the shadow flies, so flies both the year and the age like a silent army."

THERE WERE MANY influences at work in Kircher's early life, and many that had to do with the cultural rebirth for which the Renaissance is named. That revival was of course fueled by the rediscovery of ancient Latin and Greek texts, which helped to put a new emphasis on the individual, the worldly life, the civic life, and the appreciation of beauty as an important aspect of God's creation. But there was also a mystical component to it that might have had something to do with Kircher's new attitude.

In fifteenth-century Florence, in addition to commissioning work from the great Italian painters, the Medicis commissioned many translations of recovered Greek manuscripts from Italian scholars such as Marsilio Ficino. Ficino gave Europe access to the works of Plato for the first time in a thousand years, as well as to those of the so-called Neoplatonists of the second and third centuries.

More important to Ficino and to his patron, he translated a whole set of newly discovered manuscripts that were thought to be much, much older, and to have been written by an ancient Egyptian named Hermes Trismegistus. What Ficino didn't know, and what no one else knew at the time, was that they were not really the work of an ancient Egyptian, or any single person. Hermes was a purely mythological figure. His dialogues were really Neoplatonic tracts written by various authors in the first few centuries after Christ. But what historian Frances Yates called the "huge illusion of their vast antiquity" had a major influence on the thinking of the time.

Ficino's translations of Hermes were published in as many as twenty editions by the time Kircher got to them, which he undoubtedly did (perhaps even while he was in Cologne). Even Copernicus helped to back up his arguments for a sun-centered universe with a reference to this fictional authority. "In the middle of all these things sits the sun," Copernicus explained, referring to the planets and the stars. After all, how could you "place such a lantern in a more suitable place, where it can illuminate everything at the same time? Certain men not foolishly call this the light of the world, others the mind, and still others the director of the universe. Trismegistus says it is God made manifest."

He was called "Trismegistus" because as a great priest, philosopher, and king, he was "thrice great." Other texts attributed to him had been known since the days of the early Church; concerned largely with astral energies and hidden sympathies between natural phenomena, they were also thought to contain prophetic statements about the coming of Christianity. Hermes—Mercurius to the Romans, or Mercury—was said, despite his pagan status, to have had access to a secret tradition of original wisdom, and to have lived around the time of Moses.

The exact chronology was unclear, to say the least. Saint Augustine wrote, for instance, that Hermes lived "long before the sages and philosophers of Greece, but after Abraham, Isaac, Jacob, Joseph, yea, and Moses also: for at the time when Moses was born, was Atlas, Prometheus's brother, a great astronomer, living, and he was grandfather by the mother's side to the elder Mercury, who begat the father of this Trismegistus."

For his part, Ficino sometimes said Hermes was the same person as Zoroaster, or Zarathustra, known today as a prophet of Persia, and sometimes said that Zoroaster had been his predecessor. Either way, in his preface to the translations he called Hermes "the first author of theology" and traced a line of teaching that went almost directly from Hermes to Orpheus to Pythagoras to Plato. That is, according to Ficino, Plato based his philosophy on the wisdom of Hermes, who based his on even earlier knowledge. "Hence there is one ancient theology," Ficino wrote, one original theology to which the one true Christian religion was an entirely compatible heir. Hermes "foresaw the ruin of the antique religion, the rise of the new faith, the coming of Christ, the judgment to come, the resurrection of the world, the glory of the blessed, and the torments of the damned."

The dialogues attributed to Hermes are a conglomeration of mystical philosophy and poetic religiosity. But along the lines of the actual works of Plato and the Neoplatonists, they asserted that matter, the physical world, the body, was a kind of unreality. Reality, truth itself, was in the immaterial essence of things, and existed as an emanation or divine force from God. On one hand, each physical thing or phenomenon was a limited, more or less vulgar incarnation of an unlimited, pure idea. On the other hand, this immaterial truth was in all things; all matter had the immaterial energy of God within it. And though human beings lived in physical bodies, they were at least partly divine, because they had immaterial intellect. As stated by the nonexistent Hermes, this godly intellect allowed man to move through the material world "as though he were himself a god," tapping into the subtle energies that continually streamed down from above, and engaging with the invisible interconnections among stars, plants, stones, and animals.

Ficino became very involved with this *magia naturalis*, or natural magic—especially in the use of talismans to draw down astral power and in the use of music as incantation. Sometime later, Giovanni Pico della Mirandola, one of Ficino's fellow scholars, wrote his *Oration on the Dignity of Man*. "If rational," he wrote, man "will grow into a heavenly being. If intellectual, he will be an angel and the son of God." The oration is often referred to as a manifesto of worldly Renaissance humanism, a document that helped put the focus on human capacity rather than the spiritual afterlife. But it was written to introduce Pico's nine hundred theses, largely a compendium of mysticism and magic by which he believed it was possible to grasp "everything knowable."

Pico gave short shrift to Aristotle, but embraced the philosophy of the Platonists, the numerology of the Pythagoreans, the oracular poetry of the Chaldeans, the Hymns of Orpheus, the astral magic of Hermes Trismegistus, and, most especially, the Kabbalah of the Hebrews. Kabbalah was compatible with Christianity, too. One of Pico's sections was titled "[Seventy-two] Cabalistic Conclusions According to My Own Opinion, Strongly Confirming the Christian Religion Using the Hebrew Wisemen's Own Principles."

For Pico, magic was "the practical part of natural science," and also the "noblest part." It was within this context that learned people began to consider all manner of magical and mystical texts. The authority that this fake Hermes lent to Plato helped put a general premium on the immaterial: invisible bonds, unseen correspondences within the natural world. These things were "occult," but not in the modern sense; *occult* originally meant just "hidden," "concealed," or "secret."

And so a century later it was possible to be an Aristotelian, Catholic, Hermetic, and generally mystical mathematician who believed that it just might be your calling to achieve a universal understanding of things, to uncover or to *dis*cover hidden truths. After all, Hermes instructed you to "believe that nothing is impossible for you, think yourself immortal and capable of understanding all, all arts, all sciences, the nature of every living being."

Notwithstanding the Vatican's censure of various kinds of magic and strict adherence to Aristotle in natural philosophy, the Jesuits contemplated these ideas as well—though as one Jesuit philosopher wrote, "Scarcely any mortal or certainly very few indeed, and those men of the keenest mind who have employed diligent observation for a long time, can attain to such natural magic."

Young Kircher believed he might be one of these very few.

4

Scenic Proceedings

As part of their preparation for lives of obedience, Jesuit scholastics were, and still are, frequently uprooted and re-assigned. Pretty soon this newly immodest young man was sent off again, to a place called Heiligenstadt, in a relatively remote part of Saxony, to teach what Kircher rather condescendingly called "the rudiments of grammar." It may be that his superiors hoped to reintroduce him to the humility they had only recently urged him to leave behind. But it was probably too late for that. As one Jesuit historian has said about Kircher's previously humble pose, "later he tended to over-compensate at times for his early behavior."

The route to Heiligenstadt—where, as it happened, Kircher's father had once been a lay instructor to Benedictine students—passed through Fulda. His father and his mother had both died since he'd been gone, and his brothers were out in various rectories and monasteries, but his sisters (their names were Agnes, Eva, and Anna Katharina) still lived in the region. One or more of them may have warned him about the rest of the way.

"I was advised by many to change my religious garb," he remembered, "since the area to be traversed was infested with heretics." But Kircher refused, saying that he would rather die in his black cassock than make his way safely in any other clothes—perhaps also thinking that the last time he'd worn secular attire, mendacious Düsseldorfers had seen fit to trick him, and he'd almost drowned in the icy Rhine.

Kircher left with a messenger as his hired guide and headed through the region of the Eichsfeld, the "field of oaks," a rustic source of some of the fairy tales later collected by the Brothers Grimm. At one point Kircher and his companion entered "a certain dark and bristling valley," as he described it in his memoir, "which from its formidable appearance had earned the name the Valley of Hell." Suddenly they were "surrounded by heretic horsemen," who focused on Kircher's robes. "Upon recognizing from my clothes that I was a Jesuit, they immediately stripped me of everything, save my undergarments," he wrote. "After I was robbed of all my clothes, traveling provisions and books, and broken down with blows and lashings to boot, they prepared my death by hanging." He was dragged between two horses to a tree.

"When I saw that they were acting in earnest, so fierce and howling in their implacable hatred of Jesuits that they had utterly resolved to kill me," he remembered, "presently, with spirit composed, knees bent on the ground, and eyes raised toward the sky with tears, I passionately entrusted myself to God and Mary, giving thanks to divine goodness, which had rendered me worthy of enduring death on behalf of His own most sacred name. As the tears copiously welled up, I felt myself replete with as great an abundance of consolation as

I had ever experienced in my life, nor any longer did fear seize me, prepared as I was to pour out life and blood for God."

According to Kircher, this display had an immediate effect on one of the soldiers. Such was the power of the one true faith over the heretical kind. This soldier then gave a speech ("What are we doing, comrades?") that persuaded the others to drop the project completely and even to give back the things they'd taken.

Kircher thanked God profusely for protecting him. On the other hand, he felt some disappointment: "The unique and so longed-for opportunity to die on behalf of His glory had been lost."

Two days later, they finally reached little Heiligenstadt with its college and its old castle, built in the tenth century by a Frankish king named Dagobert. Heiligenstadt means "Holy City." It's now called Heilbad Heiligenstadt, literally "Spa Holy City," and in the twenty-first century, people go there for rejuvenating soaks in brine.

Kircher taught Latin there, and renewed his study of languages and mathematics with what he described as "the utmost zeal." He built another sundial, on the tower of the Church of St. Mary's. He immersed himself in the literature of the Neoplatonists and the practitioners of natural magic, and there's little doubt he became familiar with a volume called *Magia Naturalis* (*Natural Magic*) by Giambattista della Porta. The author, a sixteenth-century polymath from Naples, had written plays, made optical devices, designed military fortifications, and collected, as he claimed, "more than 2,000 secrets of medicine, and other wonderful things." The book, published in many different editions and languages over a number of decades, served as a guide on everything from cooking to the occult correspondences within nature.

"The Wolf is afraid of the Urchin," della Porta explained in one section. "Thence, if we wash our mouth and throats with Urchin's blood, it will make our voice shrill, though before it was hoarse and dull like a Wolf's voice. A Dog and a Wolf are at great enmity. And therefore a Wolf skin put upon anyone that is bitten by a mad Dog assuages the swelling of the Humor. A Hawk is a deadly enemy to Pigeons, but is defended by the Kestrel, which the Hawk cannot abide either to hear or see. And this the Pigeons know well enough."

Natural Magic went into detail on the spontaneous generation of small creatures from various forms of putrefaction—specifying what kind of dung produced which kind of insect, for example. It also told readers how to hybridize flowers and preserve fruit, distill oils and essences, extract tinctures, breed dogs, lure animals, tenderize meat, temper steel, write with invisible ink, and send secret messages. It covered medicines and remedies for common wounds, poison, and the pox, plus ways to engender sleep and different kinds of dreams. Della Porta knew how to put "a Man out of his senses for a day." And he spent thirty chapters on how "to Adorn Women, and Make them Beautiful," including how to dye hair, remove hair, curl hair, and "take away Sores and Worms that spoil hair."

Readers like Kircher learned in *Natural Magic* about various "experiments" related to light and heavy bodies, wind, air, music, and sound. Della Porta devoted many chapters to "Looking-Glasses," spectacles, and lenses—some that could be used to project "diverse apparitions of images," others to "see very far, beyond imagination." (In the years before he died in 1615, della Porta even claimed, with some reason, to have invented the telescope.) There were fifty-six chapters on "the Wonders of the Lode-stone," or the magnet. And there was a big section on "Artificial Fires," including "Fire-compositions for Festival days" made from potassium nitrate, sulfur, and charcoal, which Kircher put to use there soon enough.

AROUND THIS TIME, the region of Eichsfeld came under the jurisdiction of the Archbishop and Prince-Elector of Mainz, Johann Schweikhard von Kronberg. As the head of the electoral college that chose each new emperor, he was the most powerful Catholic official in German lands. He planned to send a formal embassy of representatives to Heiligenstadt, which in turn prepared to greet the officials with a proper reception and entertainment. As Kircher recalled, "A magnificence not to be scoffed at was being deemed appropriate for rightly receiving this group." Twenty-four-year-old Kircher was apparently so well versed in the natural magic of della Porta and others that he took charge of the "scenic proceedings."

On the evening of the event, he produced "optical illusions on a grand scale as well as a pyrotechnic display" for the visiting dignitaries, sending "fiery globes" and "shooting stars" sailing through the night sky. Most dramatic of all: an illuminated flying dragon. In general, as Kircher himself recalled, "I was exhibiting things which seemed to smack of something beyond the ordinary."

Among most of the members of his audience, these things were "stirring up the greatest admiration," as he put it. But they were causing other reactions too: "Several accused me falsely of the charge of magic," he remembered.

"Magic" was a loaded and changeable term in the early seventeenth century. As the title of della Porta's book suggests, his magic was thoroughly natural, even if some of it was "occult," since the natural world was full of concealed features. But the ability to manipulate these properties was hardly commonplace, and for some, a flying dragon or "artificial fire" signaled communion with "bad angels"—demonic magic.

"In order to free myself from this lowly charge," Kircher remembered, "I was forced to reveal for these legates the methods and knowledge behind the display. I satisfied this request to their utmost and complete satisfaction; indeed, from that time on I was barely able to separate myself from them."

He may have feigned irritation, but the "lowly charge" against him was a form of compliment. Essentially they had taken his bait. And Kircher capitalized on the opportunity of their interest to show them some "new discoveries of curiosities of mathematics" and to present them with a "panegyric of exotic languages which bore a written dedication of praise to them." This panegyric wasn't something he just had lying around. He'd worked it up to impress them. In the end, according to Kircher, "those men departed completely satisfied in every way." And when they returned to the court of the archbishop and prince-elector, they "noised about to such a degree concerning my trifles that the Prince was struck with the greatest desire to meet me."

5

Chief Inciter of Action

The Prince-Elector of Mainz was old, and in engravings he looked portly and somewhat paranoid. He'd recently built himself an immense new second residence on a high embankment above the Main River in Aschaffenburg—about fifty miles from Mainz itself. Made out of red sandstone, with five stout towers, six hundred windows, and a moat, it was part palace and part fortress. Some historians say that the assets of people killed in the ongoing hunt for witches helped fund the construction. At least fifteen hundred people were executed during Kronberg's twenty-two years as elector, a statistic that puts the legates' initial charges of demonic magic into sobering context. And now Kircher was there essentially to entertain him. But he must have succeeded, because he earned himself a place in the elector's court.

Once in residence at Aschaffenburg, Kircher devoted himself to the "private recreation" of the elector—"wholly occupied with exhibiting to him those curiosities in which he was so greatly delighting." These included a mysterious clock that Kircher said was powered by a sunflower seed, and a trick in which a small figure of Christ walked on water and saved a figure of Saint Peter from drowning. "When a strong magnet is placed in Peter's breast," he later wrote by way of instruction, "and with Christ's outstretched hands or any part of his toga turned toward Peter, made of fine steel, you will have everything required to exhibit the story. With their lower limbs well propped up on corks so that they don't totter about above the water, the statues are placed in a basin filled up to the top with water, and the iron hands of Christ soon feel the magnetic power diffused from the breast of Peter."

In the seventeenth century, even a simple magnetic trick like this had the potential to impress: here was true natural magic. As opposed to astral influence, devil incantation, godly intervention, and other invisible forces whose existence could only be assumed, magnetism, an invisible and apparently immaterial power, produced very real, reliable effects on the material world. It was believed by many to function, on earth and everywhere, almost as a living spirit.

As a thirteenth-century tract had it, for example, the lodestone "restores husbands to wives and increases elegance and charm in speech." It also cured "dropsy, spleen, fox mange, and burn." Magnetic plasters, made from shavings of iron or lodestone, were commonly applied to the body to draw out ill humors; magnets themselves were swallowed to draw them up from within. A kind of magnetic attraction, or sympathy, was also assumed to be behind the widely accepted healing action of weapon salve, used to treat men wounded on the battlefield. To make it, blood or tissue of the victim was mixed into the salve and then applied—to the *weapon* that had injured him. It was supposed to heal the wound from almost any distance. For his part, the Renaissance magus Paracelsus had promulgated the notion that disease and illness could be transferred "magnetically" to a lower life-form—that gout, for instance, could be drawn away by taking the afflicted person's toenails and implanting them in the trunk of a tree.

In 1600 a physician in the court of Queen Elizabeth of England published what is often called the first real work of experimental science, on this same subject of magnetism. In the Latin text of *De Magnete* (*On the Magnet*), William Gilbert explained how he systematically tested and debunked many commonly held notions about the lodestone. It wasn't true, for instance, that if a magnet was "anointed with garlic" it ceased to attract iron. To check the claim made by della Porta in *Natural Magic* that diamonds can magnetize iron, Gilbert conducted "an experiment with seventy excellent diamonds, in the presence of many witnesses, on a large number of spikes and wires, with the most careful precautions." It didn't work.

Gilbert's investigations resulted in a great deal of real information about the actual properties of magnets and how they behaved. When it came to questions about what magnetism *was* or how it worked, however, he took a more spiritual turn. (In some senses it was an Aristotelian-sounding turn, but it was made in the direction of Copernicus.) Gilbert concluded that a "stupendous implanted vigour"—"very like a soul"—was responsible for magnetic action and attraction, and that the earth itself was a giant magnet. He used a magnetic sphere he called a terrella, a "little earth," to perform his experiments. When the terrella was set at an angle toward the plane of another magnet, for example, it rotated. Gilbert believed the sun, "the chief inciter of action in nature," brought about the rotation of Earth in a similar way: Earth's "astral magnetic mind" responded when the sun sent forth its living energies, rotating steadily for uniform access to the vitality of its rays.

This idea influenced Kepler, who wrote that he "built all Astronomy" on the work of Copernicus, Brahe, and Gilbert, and adopted the notion of a sun that emanated a magnetic force, causing the planets to move. Magnetism seemed to explain why the planets traveled in elliptical orbits, as he'd correctly calculated. According to Kepler, "the variety of all planetary motions derives from a very simple magnetic force just as all the motions of a clock derive from a simple weight." Galileo was also influenced by the idea, and used the analogy of magnetism to explain why Earth held its axis through daily and annual motions.

Kircher's own work with magnetism extended beyond parlor tricks. In order to chart a portion of the elector's territory (newly "restored" to him through politics associated with the larger upheaval of the Thirty Years War), Kircher invented—or rather, claimed to invent, as it was subsequently revealed—a cartographic instrument that integrated a magnetic compass with measurement and drafting tools. He called it the *pantometrum*, or pantometer, a name that suggested it "measured all things." It was later described as a device for calculating "length, breadth, heights, depths, areas, of both earthly and heavenly bodies, etc."

Kircher finished the survey quickly, and the elector was "delighted to such a marvelous degree" that he "commanded that the other disputed states of the Archbishopric . . . be charted with like diligence." But the elderly elector died about a year into Kircher's service, and his successor, one Georg Friedrich von Greiffenklau, either was unimpressed with Kircher or didn't require his services, so he was assigned to what must have felt like square one, the Jesuit college at Mainz, site of his hernia-inducing skating accident, to resume the regular path to ordination.

DURING TWO YEARS of teaching in Mainz to complete what is known as the Jesuit regency period, Kircher did what he could to satisfy his now impossible curiosity and to make headway in his ongoing pursuit of the divine mind. He began to make his own night-time observations with a telescope, or a "celestial tube," as he called it. And in order to observe the sun without staring directly at it, he used a device called a helioscope, an innovation of the Jesuit Christopher Scheiner that combined telescopic lenses with mirrors to project the image of the sun onto paper or a screen. Whether Kircher built his own isn't clear, but he claimed that on a certain day in April of 1625 he witnessed for himself what Scheiner and Galileo were arguing about, observing twelve major and thirty-eight minor sunspots. It was "not without wonderment," he wrote later, that he saw "the whole heterogeneous surface of the solar hemisphere, appearing composed out of shadows and little lights."

When the Society decided to keep Kircher in Mainz for his three-year course in theology, probably because of the war, he made the most of it. "I was utterly occupied with this one endeavor," he remembered, "namely that I link to my theological studies the study of oriental languages, and that I pore equally over each at all times." In his search for the earliest Christian scripture and the ancient, divinatory theology of Hermes, Orpheus, Maimonides, Zoroaster, and others, Kircher expanded his study of languages beyond Latin, Greek, and Hebrew to Arabic and two forms of what is now called Aramaic: "biblical," or Chaldean, written in Hebrew characters, and "Christian" written in the Syriac alphabet.

But by 1629, after being kept about five years in Mainz, Kircher grew dispirited, as would any melancholic with ambitions of grandeur. (The city had been a site of frustration for inventive, ambitious sorts before; it was where Johannes Gutenberg first employed movable type, printing one hundred eighty copies of the forty-two-line Bible before being sued by his creditor and forced to stop.) Although the war had temporarily subsided, it had dragged on for a decade, politics across the Continent had grown more intricate, and almost all the nation-states of Europe had gotten involved in one way or another. Towns and villages throughout the so-called empire had been made vulnerable to the desperate brutality, not to mention the smallpox and typhus, of ill-fed armies. The plague had spread through that part of Europe as well; at its worst, in Prague, it wiped out sixteen thousand people. Harvests, years of them, had been ruined. Peasants had revolted by the thousands. The only piece of good news, from Kircher's point of view, was that after defeat in battle and illness from the campaign, Christian of Brunswick, the Insane Bishop, had died a few years before. He was twenty-six. People said that his insides had been eaten away by a huge worm.

The prospects across the German provinces were generally bleak. In January of that year, Kircher brazenly addressed a letter to the superior general of the Jesuits in Rome, making a vehement plea to be sent *somewhere* as a missionary. He was finally about to be ordained into the priesthood, and he was willing to go to just about any corner of the world to propagate the faith—"Arabia, Palestine, Constantinople, Persia, India, China, Japan, America." But he expressed a clear preference for the Holy Lands and North Africa, places where in his off-hours from saving souls he might dig up ancient scrolls and texts containing early mystical wisdom. "For the love of God, and the holy Virgin Mother," he wrote, "I resolutely implore and beseech you to grant my extremely great desire to follow the apostolic pursuit. May my prayers and supplications not be made in vain, I pray—do not permit my soul to waste away cramped among the confines of this barren Germany. Stretch forth my soul, heretofore enchained, now entirely in the service of extending the divine majesty."

"Life is short," Kircher reminded the superior general, and he certainly didn't want to spend the rest of his in Mainz.

This wasn't exactly the humility and indifference that Ignatius of Loyola had in mind for his soldiers of God, the kind that meant you "do not desire, nor even prefer" one circumstance over another as long as they served God equally. The Jesuit authorities did not grant his request. Instead, after Kircher's ordination they reassigned him to another old city on the Rhine. This time Speyer, where he spent a customary period of spiritual probation before saying his final vows.

ONE DAY IN SPEYER, Kircher was asked by a superior to find a book in the library. While looking among the stacks—"Was it by chance or by the arrangement of divine providence?" he later wondered—he came across a volume depicting several ancient Egyptian obelisks. These particular obelisks, now in Rome, were thought to have been brought back from Egypt by conquering generals as many as fifteen or sixteen hundred years prior. Fallen into ruins over the centuries, they were restored and re-erected by a recent pope in the decade or two before Kircher was born.

"Instantly carried away with curiosity," Kircher assumed for a moment that the hieroglyphic markings on these structures were artistic decorations. But "when from the attached history of obelisks I learned that these figures were the chronicles of ancient Egyptian

Wisdom, inscribed from time immemorial . . ." he recalled, "the desire befell me and I was goaded by the greatest hidden impulse to discover whether it was possible to attain the acquisition of knowledge of this type."

After all, an explanation of these markings "had been offered by no one since their meaning had been destroyed over the passage of so much time." Many believed that Hermes Trismegistus himself had devised the hieroglyphs as a way of preserving and protecting the old wisdom, encoding it in symbolic language that was universal but also indecipherable to everyone but the truly wise. "It was the opinion of the ancient theologians," wrote Pico della Mirandola, "that one should not rashly make public the secret mysteries of theology." The obelisks were thought to contain some of the earliest and most sacred ideas of all: possibly this was a strain of knowledge that originated in the time of Adam, a strain that had survived the Flood and the confusion of tongues.

"From that very moment I never turned my mind from deciphering these figures," Kircher claimed. "For I was reasoning thus: imprinted characters of the ancient Egyptians have survived, indeed even genuine ones at that; therefore, the meanings of these characters will still somewhere lie hidden, scattered among the chronicles of ancient authors, and perhaps not in Latin and Greek texts but in those exotic works of the Orient."

Later that year, about a decade after arriving on gangrenous feet for his novitiate in Paderborn, Kircher made his final vows as a Jesuit priest—retaking vows of poverty, chastity, and obedience as "perpetual solemn vows," and making an additional vow of obedience to the pope. For his first assignment as a fully professed priest,

Kircher headed back up the Rhine, past Mainz, back up the Main, past Aschaffenburg, into a region of centuries-old vineyards, to a university town called Würzburg.

BY THE TIME Kircher arrived in 1630, Catholic victories against Danish Protestants in the war had resulted in a peace treaty. For the time being anyway, at least according to Kircher, "high peace resided over the Catholics." On the other hand, the witch hunt that had been going on for the last several years in the archbishopric of Würzburg wasn't quite over. From his fortress on a high slope across the river from the city, the prince-bishop had overseen an investigation in which as many as nine hundred people were executed, including members of the clergy, many young children, and his own nephew.

Around this time, a Jesuit named Friedrich Spee, long believed to have been a confessor to the condemned in Würzburg, wrote anonymously against the persecutions and described the "wretched plight" of someone who had been tortured into confessing her guilt. "Not only is there in general no door for her escape," he wrote, "but she is also compelled to accuse others, of whom she knows no ill, and whose names are not seldom suggested to her by her examiners or by the executioner. . . . These in their turn are forced to accuse others, and these still others, and so it goes on: who can help seeing that it must go on without end?"

Kircher doesn't mention the witch hunts in his memoir. Against this dark backdrop he is known to have taught mathematics, philosophy, Hebrew, and Syriac, and to have built two new sundials, on the south and the east sides of the university's central tower. He also developed a very close friendship with a younger, awestruck student named Kaspar Schott, with whom he apparently composed music. Neither Kircher nor Schott could have foreseen how they would be separated and reunited and separated again in the years to come.

Kircher wrote his first book manuscript in Würzburg, although at only sixty-three pages *Ars Magnesia* (*The Magnetic Art*) was more like a pamphlet, and since modern scholars see it as "highly derivative" of Gilbert's already famous work on the subject, perhaps it wasn't entirely *his*. Kircher steered clear of Gilbert's heliocentric ideas but echoed his views on magnetic attraction, describing it as "a primary and radical vigor." And he agreed that the earth behaved somewhat like a magnet: things are drawn down toward the earth, he suggested, putting his Aristotle on display, like the natural attraction of something to that which is good for it. The entire second part of the book, however, was given over to practical and recreational uses of the lodestone, something Gilbert hadn't really bothered with, including instructions for the trick in which Christ rescues Saint Peter from drowning.

But the "high peace" that Kircher described, such as it was, wasn't meant to last. The Thirty Years War was only in its twelfth or thirteenth year, and soon "new and sudden whirlwinds of wars rendered all things topsy-turvy." The king of Sweden, Gustavus Adolphus, had taken up the Protestant cause, defeating the Catholic general, the count of Tilly, at a place called Breitenfeld. Now he was marching his armies rapidly through central Germany and headed in their direction.

At Würzburg, "the entire College was dissolved within twenty-four hours of unbelievable confusion," Kircher remembered. "All were shaken by terror as the enemy was now arriving at the city; for they had heard that they would spare not one of the Jesuits. I, too, rolled in this communal whirlwind." He left the city with others for Mainz, abandoning the pages of a new manuscript.

After four days of siege in mid-October, Gustavus took Würzburg. He took the city of Hanau a few weeks later, Aschaffenburg a number of days after that, Frankfurt less than a week after that, and Mainz five days before Christmas. Kircher was separated from his friend Schott somewhere along the way, and fled again—back to Speyer, and then out of Germany altogether, leaving behind an entire region (for good) like a devastated home.

"At Bamberg the bodies lay unburied in the streets, and on both sides of the Rhine there was famine," C. V. Wedgwood wrote about the eventual aftermath of the Gustavus campaign. "In Bavaria there was neither corn left to grind nor seed to sow for the year to come; plague and famine wiped out whole villages, mad dogs attacked their masters, and the authorities posted men with guns to shoot the raving victims before they could contaminate their fellows; hungry wolves abandoned the woods and mountains to roam through the deserted hamlets, devouring the dying and the dead."

6

Beautiful Reports

S ince all things in Germany had been turned upside down, and since there shone no hope either of remaining or of returning," Kircher and others were sent to France. They traveled down the Rhône valley to Lyon, where there was a Jesuit school—but also unfortunately where there had been another outbreak of the plague. So he was sent farther south, to Avignon, which must have seemed like a different world.

France's own wars of religion were over, for the time being. Although Louis XIII had secretly and then not so secretly allied himself with the Protestants against his Hapsburg enemies in Germany, he and his chief minister, Cardinal Richelieu, had pretty well driven out the heresy of the French Protestants, the Huguenots, within his own territories. France was Catholic and Avignon itself, where a number of French popes lived during the fourteenth century, was still a papal territory. There were so many bells in so many steeples in Avignon that it was known as *la ville sonnante*, "the ringing town," and it's easy to imagine that for Kircher a feeling of security, rather than alarm, began to accompany the sound of their pealing.

A new Jesuit college was being built there out of white sandstone, around a large square and gardens, with tall portico archways and large windows above. Many of the buildings in Avignon—the Palais des Papes, Notre Dame des Doms, the Pont d'Avignon—were made out of the same stone, which has a way of taking on the color of the day. During parts of the year the infamous mistral winds blow cold down the Rhône valley on Avignon, but also blow the cloud covering away, leaving crisp air, the warmth of the sun, and the blue, as Kircher described it, of "an Egyptian sky."

It's hard to say how well this twenty-nine-year-old priest from war-torn Germany was received in the south of France. As records from the college at Avignon show, Kircher's superiors thought his "talent" was "good," that his "accomplishment in letters" was "great," and that his "ministry" should be "teaching"—but that he had only "some" "discretion," and that his "experience of things," by which they seem to mean his level of maturity, was "not great." Despite this less than enthusiastic assessment, Kircher went on with what has been called his "strange combination of mathematics and biblical languages." When he wasn't teaching or studying, he was up in the college tower, working on a project inspired by the Avignon light.

Kircher set up mirrors at the windows that reflected the sunlight onto the tower's arched ceiling and walls, where it traced a path across marked astronomical points, constellations, and astrological signs. It was a little like a planetarium, or upside-down sundial, that also indicated the time of day in different locations around the world and helped chart horoscopes. The project became the basis for Kircher's next book, which was printed a few years later.

He took up direct observation of the sky as well. All over Europe, educated men (because, again, not many women were given educations) with a sense of curiosity and the money or craftsmanship required to own a telescope were trying to see for themselves what all the fuss was about—why their entire understanding of the universe might have to change. The south of France in particular was already known as a good place to use the astronomical tube. While he was there, Kircher had contact with, and tried to convert, an astronomer and Hebrew scholar named Rabbi Salomon Azubins de Tarascon. He made celestial readings with a traveling student from Danzig named Johannes Hevelius, who was more interested in telescopes than in his family's brewing business. And while traveling near Aix-en-Provence in the fall of 1632, as Kircher put it, he "fell in with"—or made a point of falling in with—someone who was not merely an astronomer but "the most celebrated man, the greatest patron of letters in all of Europe, a Senator in the Parliament there, Nicolas Claude Fabri de Peiresc."

Kircher's assessment of Peiresc's status was not far off. The son of a wealthy magistrate with connections to the royal court, Peiresc did occupy his family's seat in the parliament of Provence, and he was an enthusiastic champion of talented scholars. As a young man, after studying with the Jesuits and training as a lawyer, he traveled around for a while, making associations within Europe's intellectual and cultural circles. Then he went home to fulfill his political duty, live out the ideals of the Renaissance humanist, and support the general advancement of learning. Dividing his time between his home in Aix and his family's country estate in Belgentier, near Toulon, Peiresc studied old manuscripts and collected coins, paintings, antiquities, natural curiosities, and zoological specimens. He observed the moons of Jupiter for himself not long after Galileo discovered them, and recorded with precise notation the first sighting of the Orion

Nebula. At Belgentier, in addition to growing malvoisie grapes for the bottling of his own wine, he cultivated sixty types of apple, twenty varieties of citron, a dozen sorts of orange, and all kinds of melons, apricots, and olives.

But Peiresc spent most of his time writing letters. According to his protégé, the mathematician and philosopher Pierre Gassendi, "On those dayes on which the Posts did set forth towards Paris or Rome, he was wont to defer his Supper, till ten or eleven a Clock, and very often, till after mid-night; that he might write more, and larger letters." After his death, a niece is alleged to have burned some portion of Peiresc's hand-copied correspondence for heat; even so, *ten thousand* of his letters have survived.

Peiresc functioned like a hub in a virtual network that came to be called the Republic of Letters, continually exchanging intellectual news and information with a wide circle of scholars, philosophers, and artists across Europe and beyond. His elite status meant that he could be in frequent correspondence with people such as the painter Peter Paul Rubens, the cultural patron and bibliophile Cardinal Francesco Barberini, and Francesco's uncle Maffeo, who in 1623 became Pope Urban VIII. Anything interesting that he sent to his friend Marin Mersenne—a Minim friar in Paris who wrote about theology, math, and music—might be forwarded to, say, the English philosopher Thomas Hobbes, or the lawyer and amateur mathematician Pierre de Fermat, who might send it off to someone else.

As a modern scholar has described it, the first meeting between Peiresc and Kircher "could not have been very intimate," since in subsequent letters to others, Peiresc referred to his new acquaintance variously as Balthazar Kilner, Balthazard Kyrner, and Athanase Kirser. But Peiresc was intrigued. Kircher told him that he'd been working on a Latin translation of a rare Arabic manuscript that he had saved from the prince-elector's library at Mainz before the heretic armies came through. As Peiresc later reported, the document was supposedly written by "Rabbi Barachias Nephi of Babylon" and offered insight into "interpreting and deciphering the hieroglyphic letters" of the ancient Egyptians.

Like Kircher and many others, Peiresc was fascinated by the notion that the hieroglyphic texts represented very early learning. According to Gassendi, Peiresc's most prized library possession was a papyrus scroll "all written with Hieroglyphick Letters" that had been "found in a Box at the feet of a certain Mumie." Peiresc theorized that Coptic, the language of the early Egyptian Christians, might shed light on the workings of the hieroglyphic system. He'd been corresponding with a pair of Capuchin monks in Egypt and working through agents in Cairo to purchase anything he could that was in Coptic or about Coptic.

Several months after he and Peiresc first met, Kircher sent him a sample from his translation of the mysterious Arabic text. Peiresc wrote that it "made me more hopeful than I used to be of the discovery of things that have been so unknown to Christianity for close to two thousand years." When the next sample came, he wasn't *quite* as effusive, but urged Kircher to come for an extended visit and to bring the manuscript with him for Peiresc to see. Kircher was uncharacteristically slow to accept the invitation of such an important person, but finally went to stay for four or five days in the spring of 1633.

ALTHOUGH THIN and prone to sickness, fifty-two-year-old Peiresc was a man of "rare courtesie and affability." During Kircher's visit

Peiresc wrote that he and Gassendi were having "great pleasure in taking him around." It wasn't just the three of them: Peiresc kept dozens of cats "by reason of mice," and when he went outdoors, a servant "waited upon him with an Hand-Canopy, to keep off the Sun-beams." In his gardens and sunrooms he grew black locust from America, jasmine from Persia, ginger from India, rare vines from Damascus, and one China rose. He owned five telescopes and a large treasure of books, with which Kircher demonstrated his facility with languages.

There was plenty to discuss (they probably spoke Latin), and these French were sophisticated: They took an empirical, mathematical approach to the study of nature and had pretty much given up on Aristotle. They were generally skeptical, especially when it came to fantastic explanations for physical phenomena. Peiresc had withheld judgment some years before, for example, when the bones of a giant were discovered in an ancient tomb; many years later he was able to determine that they actually belonged to an elephant.

Gassendi and Mersenne had recently been involved in a public feud with a physician, alchemist, astrologist, cosmologist, and Kabbalist named Robert Fludd. Fludd believed in the idea of a "world soul," invisible harmony between microcosm and macrocosm, the divinity of Hermes Trismegistus, and the ability of the weapon salve to heal wounds from a distance. Mersenne, remembered today for the prime numbers named after him, wrote that Fludd was "an evil magician, a doctor and propagator of foul and horrendous magic." Gassendi had been a little more gracious. A devoutly pious priest, he happened to believe that everything was composed of basic units of matter called atoms—a material theory he got from his own ancient sources, Epicurus and Lucretius.

On some of these topics, Kircher kept his mouth shut—on others, not as much. Fludd's ideas were closer to his own than he might have been willing to admit. But after more than a decade of rigorous scholarship, Kircher could impress with his erudition, and his storytelling had a disarming effect on people. In the end, the skeptics were taken with him. "He has beautiful reports and beautiful secrets of nature," Peiresc said in a letter. Kircher told them about his sunflower-seed clock, and it sounded "very marvelous" to Peiresc, who added: "He promises to show us proof."

As Kircher described it, the clock told time because the sunflower seed always turned toward the sun, the way the flowers do, by virtue of magnetic attraction. "And he says he has shown proof of it in good company at the dinner table, in the presence of the Elector of Mainz," Peiresc wrote, "and even if one was in the house and out of the sunlight or if the sky was covered with clouds, the clock would never stop showing the most precise time, as much as is possible in the arc that the sun makes in our horizon. Even if it weren't so exactly precise, and if it showed no other change than turning successively at sunrise or at noon or at sunset, more or less, I would still hold it as a great miracle of nature and one that well deserves to be seen."

Kircher may have intended the clock to suggest the degree to which invisible forces—of the kind that Fludd endorsed—were natural enough. But other implications weren't lost on Peiresc. That same month, Galileo was standing trial in Rome. (His *Dialogue Concerning the Two Chief World Systems* had not only espoused the Copernican model but also made some fun of arguments for an Earth-centered universe and, at least it appeared, of the pope.) Peiresc understood that evidence of the sun's magnetic action or attraction could help make the case for Earth's rotation and revolution. After Kircher's visit, Peiresc wrote about the sunflower-seed clock to Marin Mersenne in Paris, who wrote about it to, among others, René Descartes.

"If the experiment that you describe to me of a clock without sun is certain, it is quite curious, and I thank you for having written to me about it," Descartes replied from Deventer in the Netherlands. "But I still have doubts about the effect; even so, I don't judge it impossible."

Strangely, the original purpose of Kircher's visit had to be postponed. When the time came to begin a discussion about hieroglyphics, it turned out that Kircher had failed to bring along the Barachias Nephi manuscript. This is probably because, to one degree or another, he'd exaggerated its very being. As historian Daniel Stolzenberg writes in his doctoral dissertation on Kircher's Egyptological efforts, "No such author or text is known to exist." Kircher didn't make it up entirely—perhaps there was a compilation of old writings—but it wasn't quite what he said it was.

Kircher left with a trunk-load of books Peiresc had given him to help with his translation. In return, he promised to come back again to demonstrate the clock and to be sure to bring the manuscript with him when he did.

SOON AFTER RETURNING to Avignon, Kircher received a letter from the superior general of the Jesuits containing news about another reassignment. In his memoir, Kircher rendered it as if taking it in for the first time: "I am called to Vienna, Austria, where I have been designated Mathematician of Caesar."

The statement sounds almost delusional. But it seemed that "Caesar," by whom he meant Ferdinand II, the Holy Roman Emperor, was looking for a replacement for Johannes Kepler, who had died a couple of years before. (Kepler had his share of early-modern-age trouble. His mother had been accused of witchcraft—he'd used his influence to have her released. Six out of the eleven children from his two marriages died in infancy or childhood. Now, just a few years after his death, his bones were said to be lost; the churchyard in which he was buried was blasted and trampled into nothingness during the Swedish siege of Regensberg.)

The assignment was within the realm of reason: Kircher had already served at a very high level in the court at Aschaffenburg. "Upon learning this," Kircher recalled, "Peiresc left no stone unturned in his effort to impede this journey; for he was fearing that, while I was occupied with my mathematical studies in the halls of Caesar, I would distract all of my attention from attaining an understanding of Hieroglyphics."

The truth is that Peiresc wanted to keep Kircher, and the Barachias Nephi text, in Provence. He wrote to the pope's nephew, Cardinal Barberini, asking him to step in and help. If Kircher went to Vienna, Peiresc explained, his translation "will surely be delayed, and perhaps even completely confounded." At the very least, Peiresc implored in subsequent letters, it would be better to transfer him to Rome, where he would have access to the libraries of the Vatican. Meanwhile, Peiresc kept urging Kircher to return to Aix for another visit, and Kircher kept putting it off.

During this time when Peiresc was hoping for some good news from either Barberini or Kircher, both men were otherwise occupied. Barberini was busy serving as one of ten judges in the heresy proceedings against his friend Galileo. And Kircher was busy getting caught in another waterwheel.

It happened one day when Kircher needed a break from his studies and went outside to clear his head: "There was in the college a suburban garden in which a huge wheel between two walls was driven by a horse in order to irrigate the garden," he explained. "At the very bottom between these walls was a great supply of gushing water." With his mind a hundred miles away, he sat down "on the aforementioned machine, which was being driven with a bar by a huge horse." Immersed in his thoughts and "paying the toiling horse no heed," he was "suddenly snatched away by the bar, and since I was able neither to secure the horse nor to halt between the wall and the bar without the danger of entirely crushing my body, I was suddenly by the bar cast down within the wheel. But since the wheel was moving unceasingly, nowhere could I set my foot, nor was it possible to slip away from the side on account of the narrowness of the wall, which was nearly touching the wheel." He called out to a fellow Jesuit who was walking in the garden, but he didn't hear. "In the meantime," he remembered, "I was being rolled around with the wheel."

There was little question about what to do. "With my usual faith I took refuge to the Blessed Virgin," Kircher remembered. "And, lo, the wheel stopped." Kircher claimed that the event haunted him even in his old age: "This potential disaster was so formidable that I am not able to think on it without horror." But because his "escape from danger was achieved with divine aid," he claimed his resolve to serve God was reinforced "to the utmost degree."

———————

AT THE END of summer, without any apparent success on Peiresc's part to delay or change his assignment, Kircher was left with no choice but to leave for Vienna. On his way to the seaport of Marseille in early September, he stopped in Aix, finally making the return visit to Peiresc that he'd promised.

He also finally gave a demonstration of the sunflower-seed clock he'd talked so much about. Kircher required some time on his own to set it up. Peiresc the lawyer prepared to take detailed notes. When the moment came, the seed was inserted within a cork that floated in a pot of water; the hours of the day and meridian lines were indicated around the pot's rim. The relative time was indicated by a little pointer; as recorded by Peiresc, it was "one third of an hour after two in the evening or afternoon." Kircher's markings also showed "by definite relation what the time it was in Rome, Constantinople, Jerusalem, Babylon, the Indies, China, America, Peru and the Canarys, and also other places." And, sure enough, no matter which way Kircher turned the cork with the seed, it found itself back at the place where it had been, facing in the direction of the sun.

But Peiresc began to suspect that something besides attraction between the seed and the sun was at work. "What made me doubt the certitude of his experiment and of his words was the fact that he would not swear that the sunflower seed alone was sufficient for the demonstration," he wrote. "Thus, without actually saying it, he left me with the understanding that he required some other unknown ingredient that he did not wish to declare, and which I guess to be [a] magnet."

In other words, Peiresc believed that the cork contained not only a sunflower seed but also a hidden lodestone. And he was right. The clock was really a compass: for the seed to face the sun, Kircher had to know in advance what the position of the sun was, relative to magnetic north, which meant that for his clock to tell time, he had to know in advance what time it was. Peiresc did not find this parlor trick "to be a miracle of any kind."

Things didn't go any better when discussion turned to hieroglyphics. As if to prove that the Barachias Nephi treatise really existed, Kircher at long last brought out an Arabic manuscript for Peiresc to see. But he let him have a good look at only one page from the lexicon in the back. "The bother that he made over letting me transcribe a couple of entries," Peiresc recalled, "made me suspect that he feared that I would discover that it was nothing but a kind of translation of Horapollo." (Horapollo was the purported author of a well-known Greek text called *Hieroglyphics* that was found in the fifteenth century; it lays out a purely allegorical, as opposed to phonetic, scheme for translating the Egyptian texts.)

At some point Kircher also presented Peiresc with a paper describing his preliminary ideas, his "protheories," about hieroglyphic interpretation. It quoted Barachias Nephi and included a reading of the obelisk of St. John Lateran, based on an engraving he'd found in a certain book. But when Peiresc read it, it was clear that Kircher had chosen to analyze one of the more obviously unfaithful renderings of the obelisk.

"I discovered it unfitting that the figures were all imagined at the whim of the artist, like grotesque works, that didn't have anything to do with the ancient Egyptian style, and that didn't have any connection to the real hieroglyphic figures of the obelisk of Lateran," recalled Peiresc. "All of which I pointed out to him, and he finally admitted it, with much grief, since he had found such beautiful interpretations, and well-accredited ones, it seemed, of all the figures found there, or of most of them."

When Peiresc pointed out another case in which Kircher failed even to recognize what real hieroglyphics looked like, "he refused to admit it until he exhausted himself, surprised by the inaccuracy of the picture, for which he had left aside the more correct and faithful one to instead follow one which was totally conflicting and discordant with the style and antiquity that this work should give off."

According to Peiresc, Kircher was "very ashamed after all was said and done." But according to Kircher, who wrote about these events years later, Peiresc "was suffused with joy" and "spoke of my work with such a sublimity of words that, in keeping with modesty, I don't think it proper to describe here."

WITH PRESUMABLY AWKWARD good-byes exchanged, Kircher headed for Marseille, where he joined up with a few other Jesuits. Together they intended to sail more than three hundred miles along the Mediterranean, first to Genoa and then to Livorno (Leghorn), on the Tuscan coast. The route allowed them to avoid the war: from Livorno he could travel overland to Venice and finally to Vienna. When they sailed off, he wrote, "we entrusted ourselves to the sea of Marseille"—and also presumably to the captain and crew of the ship, who, having been paid, ditched the seasick travelers on a small barren island several miles from their starting point.

Kircher and his companions pooled their money and paid a fisherman to return them to Marseille, where they began their journey again, this time on a felucca, the kind of low craft with a lateen rig traditionally seen on the Mediterranean and along the Nile. Soon, however, they were forced to take safe harbor in a "deserted port for three days on account of the unrest of the season and of the sea." When they started again, so did the rough water. "The south wind began to rise and the sea swelled," Kircher remembered, but the captain "proceeded nonetheless with immense fortitude," until "the ship was no longer capable of sustaining the violence of the swells, the seas became so huge that I was not able to look upon them without horror, and all the while we were busy bilging from the ship all that water that had been tossed in by the force of the storm." The priests on board began taking confession from other passengers and from one another. Soon "darkness fell as an addition to our complement of myriad suffering."

The captain, who takes on an increasingly Odyssean cast in Kircher's telling, decided to head for the protection of a cavern along the coast "within a protraction of crags," whose "entrance was at one moment closed off by the flux of the waves, and at the next moment opened by the retraction of the very same swells. Guided by both his wholly clear plan and his Guardian Angel, who was directing his rudder with purpose, he observed the slipping of the waves, and forthwith directed the little ship toward the mountain side, where it was hurled into the cave by the force of the waves, more by the arrangement of God than by the industry of man; for at any other moment, while the entrance was filled with waves, we all would have perished dashed against the rocks." When they eventually got to the other side of the rocky Massif des Calanques and the port of Cassis, they had gone just eighteen miles, as the crow flies, from Marseille.

After that, Kircher made up time, taking only eight days to reach Genoa and then, after "acquiring another little boat for rent," sailing down the Italian coast toward Livorno. But in the words of an English scholar, Kircher's "presence alone seems to have been a certain guarantee of a storm." And he later claimed his boat was "driven by winds and tempests" off course more than one hundred seemingly impossible miles to Corsica, and then another impossible hundred or so back to the Italian mainland near Civitavecchia, the main seaport of Rome. Left with nothing "except hunger and calamity," Kircher walked to the city, forty-five miles away.

"And thus I reached Rome," he recalled, "where, to my utter surprise, I was being awaited."

PART TWO

7

Secret Exotic Matters

Kircher couldn't really have been too surprised, either to find himself in Rome or to be expected there. He'd indicated in a couple of letters that he planned to see the city and its Egyptian obelisks before going on to Vienna. He also knew, or hoped, that Peiresc's letters to Cardinal Barberini might finally succeed, and that there was still a chance he could be reassigned to Rome. And that's exactly what happened during the time he said his boat was bobbing around like a cork on the Tyrrhenian Sea. Peiresc's opinion of Kircher's talents may have changed, but not before his original endorsement had persuaded Barberini to keep him and his mysterious manuscript for himself, and to send the astronomer Christopher Scheiner to Vienna instead. But it should go without saying at this point that Kircher wasn't above feigning a little surprise when it suited him, and this story—that he just happened to end up in Rome, only to find out that he'd been reassigned to Rome—was subsequently repeated so many times that he may have eventually believed it himself.

If not genuinely astonished by the turn of events that kept him there, Kircher might have been overwhelmed by the city he walked into that day. Among possible first sense-impressions: a whiff of the Pontine Marshes, the three-hundred-square-mile swamp district to the south of the city; it was a foul-smelling source of malaria and other diseases that engineers had been trying to drain since the time of Julius Caesar. At the gates of Rome itself, he would have been stopped to make sure that he wasn't bringing in a case of the plague, say, or anything on the Index of Prohibited Books. Once he started walking Rome's unpaved streets, to paraphrase an early-nineteenth-century traveler, magnificence and filth frequently competed for his attention.

A little more than a hundred years before Kircher arrived, Rome was almost completely destroyed by an army of mercenaries that had gone unpaid too long. The city was physically wrecked, forty thousand people died or disappeared, and in the aftermath the population was down to about ten thousand. But Rome had been rebuilt many times, and its recent physical reincarnation, still under way, was also hugely symbolic. It was meant, in a sense, to prove all the heretics wrong: there was only one true religion and only one place where it could possibly reside. After more than a century of construction, parts of St. Peter's were still unfinished when Kircher showed up in 1633, but its dome, as re-envisioned by Michelangelo, had taken its place as the dominant backdrop of the city.

Currently the well-connected and well-educated Barberini family was involved in a twin program of papal nepotism and almost unprecedented cultural patronage, with the former helping to fund the latter. Projects in painting, sculpture, and especially architecture, by Rubens, Bernini, Borromini, and others, were bringing the baroque style into existence. In places like the Chiesa Nuova, the New

Church, "rare music" was "sung by eunuchs . . . accompanied by the-orbos, harpsicords and viols." A new musical form called *opera*, "work" in Italian, was being performed at the new Barberini family palazzo and at other places around the city. Bernini in particular was known for special effects in the theaters: he created artificial lights and fires, and made the sun rise with a machine. For one production, he simulated the flooding of the Tiber by sending a wall of rushing water directly toward the audience; it was diverted by stage design at the last minute.

So many people made the pilgrimage to the revitalized city that even before the end of the sixteenth century, when Montaigne traveled there from France, what bothered him most was how many other Frenchmen there were. Rome had about thirty thousand visitors a year, and every conceivable kind of human being and business—though it was especially replete with priests, nuns, artisans, merchants, bankers, prostitutes, and most visibly of all, indigents. "In Rome one sees only beggars," an Italian traveler complained, "and they are so numerous that it is impossible to walk the streets without having them around."

There were a hundred religious orders in Rome, and many of them tried to help the destitute and the sick, running orphanages and hospitals with thousand-bed wards. The well-off lived in the area around Trinità dei Monti—the Church of the Holy Trinity on the Pincian Hill—and off the Via del Corso. The Jewish population lived in an overcrowded ghetto in a low-lying area near the Tiber River that was prone to flooding. "Being environed with walls, they are locked up every night," reported an Englishman. "The Jews in Rome all wear yellow hats, live only upon brokerage and usury, very poor and despicable." In Campo de' Fiori, the market square where "horses, all kinds of corn, and other commodities" were sold, executions were also sometimes performed. The priest, mystic, poet, and philosopher Giordano Bruno was burned at the stake there thirty-some years before as punishment for heresy. Almost anywhere in Rome you could come across a brand-new palazzo or crumbling ruin, the smell of cooked cabbage or the odor of urine.

Kircher made his way through the confusing, narrow streets to the Jesuit Collegio Romano, where the previous pope, the current pope, and a host of cardinals had been educated. It wasn't far away on Via del Pie' di Marmo (Street of the Marble Foot) from the Dominican convent where Galileo was tried in the spring and from the massive rotunda of the Pantheon. In the judgment of a visitor, the front of the Collegio Romano gave "place to few for its architecture, most of its ornaments being of rich marble. It has within a noble portico and court, sustained by stately columns, as is the corridor over the portico, at the sides of which are the schools for arts and sciences." The college boasted a large garden, an elaborate library, and a multiroom apothecary for making everything from candle wax to the herbal concoctions that chaste Jesuits took to dampen sexual desire.

In his first few days in Rome, Kircher sought out Peiresc's contacts. "I paid my respects to those who in turn were greatly revivified by my arrival," he boasted. Cardinal Barberini "received me with so great a measure of kindness that he seemed to forestall the others by offering more swiftly than they the generosity of his own services to my undeserving self."

Barberini was only about five years older than Kircher, but his uncle had made him cardinal a decade before. In addition to directing foreign affairs for the papacy, he'd become one of the most influ-

The Collegio Romano

ential people in Rome, though he was also spending much more money than he could afford. He was a cultured, obsessive collector of old volumes and manuscripts, and had previously served as the cardinal-librarian of the Vatican. Kircher wrote to Peiresc about a second meeting with Barberini a few days later: "He questioned me about many things pertaining to letters," especially those "concerning the interpretation of Hieroglyphics." Barberini wanted to know "by what reasoning, by what method, by what author, by what inscriptions they are able to be extricated." He also "requested that I tell him what I knew about the secret rites of the Cabala, what benefit each holds indisputably for human affairs."

On the basis of his satisfying responses to such questions, Kircher came away with the cardinal's full support and an initial twofold assignment: to complete a Latin translation of the Barachias Nephi treatise and, as an example of Nephi's methodology, an explication of the mysterious Bembine Tablet. Also called the Table of Isis, the bronze tablet was elaborately inlaid with images of Egyptian figures, markings, and symbols. It had turned up a hundred years before, after the sack of Rome, and was thought to hold great, ancient secrets. If all this work went well, Kircher could proceed to a full set of commentaries on the obelisks of Rome.

There were at least two problems with this assignment. The first had to do with the questionable status of the Barachias Nephi text. The second had to do with the fact that, like the dialogues attributed to Hermes Trismegistus, the Bembine Tablet was actually created a few thousand years later than anyone thought. It wasn't ancient or Egyptian so much as it was rendered in an ancient Egyptian motif. But Kircher kept the first problem to himself and neither Kircher nor Barberini knew about the second.

Barberini offered financial support as well as special access to libraries, manuscripts, and scholars of esoteric languages in Rome. As Kircher reported, the cardinal ordered his personal librarian to take him "to the Vatican library and then to his own library and likewise to all the antiquities of the city of Rome, including all obelisks, pyramids, ruins, and statues, those scattered both in the city as well as here and there in the gardens of the Cardinals, all those things which might be able to be of use to me in order that I properly undertake this task."

He was no doubt also taken to other sites of interest. In the church of Santa Maria Maggiore, for example, there was a relic said to be part of the crib in which the baby Jesus had been laid. At the Palazzo Farnese there was a marble head of Christ supposedly carved from life. And at the resplendent Il Gesù, the Jesuit Church of Jesus, there was not only the body of Ignatius but the right arm of Francis Xavier, the first Jesuit missionary.

A few months later, a young priest and natural philosopher named Raffaello Maggiotti wrote to Galileo, now under house arrest in Arcetri, near Florence. "News is that there is a Jesuit in Rome who has been in the East a long time and, as well as knowing twelve languages and good geometry has brought with him very good things," the letter said. "Among them a briar that turns according to the sun and also serves as a perfect clock." Also among them, "Arabic and Chaldean manuscripts, with a copious display of hieroglyphs." According to Maggiotti, Kircher affirmed that the hieroglyphics were "made before Abraham was born and he says those scripts contain great secrets and stories."

Apparently the clock that had been a success with the Prince-Elector of Mainz and then a disappointment in Aix was working its wonders again. Or Kircher had been talking them up again. And as Paula Findlen, a modern scholar, has put it, if Kircher "did not deliberately deceive his Roman audience," he also probably didn't "disabuse them of the idea that he had actually been to the Orient." After all, he'd *wanted* to go.

NOT EVERYONE in Rome was impressed with the clock, or with Kircher, who wrote in his memoirs that God "sets limits" on individual desire for glory by way of "hounding persecution." It seems there were "men of letters" who were skeptical about him and his abilities, especially considering his age—"for indeed I was only thirty-two years old." These people, he recalled, "not only were harboring doubts concerning my credibility, but also lodged through false accusation the charge of imposter."

Although he'd secured an assignment to translate a manuscript whose contents he had at the very least exaggerated, Kircher was determined to prove these charges wrong. And so—"lest I fall in with the label of fraud to the detriment of my religion"—he threw himself into his work.

During his first couple of years in Rome, it wasn't always clear *what* Kircher was doing. He spent some of his time testing his ideas about the magnetic attraction of the sun with the mimosa and tamarind plants in Cardinal Barberini's botanical garden. Beyond that, Peiresc, whose reputation was now partly on the line for recommending him, worried that he was concentrating on something more like his "protheories" than on a straight translation of that mysterious Arabic treatise. He frequently wrote to Kircher imploring him to stick to translations rather than interpretations. Sometimes he made insinuations about the text, and about Kircher's intentions.

"I am hurt that you have so small an opinion of me that you judge that I presumptuously wish to undertake a matter of which I am ignorant," Kircher responded in one letter. "Indeed you waver about my good faith. Surely pretence, falsehood, and whatever is contrary to true and genuine sincerity are so foreign to me that I would prefer that all my scholarly labors perish, rather than commit such a crime in the Republic of Letters. . . . You may think that my works proceed from . . . vain glory and appetite for esteem, which I despise as diametrically opposed to piety; but I only undertook these matters lest I seemed to fail in my duty with respect to the talents granted to me by God in His infinite goodness."

Even if the manuscript had been exactly what he claimed, the diversity of stimulating materials in Rome would have made it difficult for Kircher to focus on the task at hand. The Vatican kept a vast collection of Near Eastern and Middle Eastern texts brought back as booty from the Crusades several hundred years before. Kircher came across Arabic manuscripts on amulets, Jewish manuscripts in Chaldean, inscriptions in Chinese and Syriac, and inscriptions in languages unknown altogether. He briefly began an entirely different and even more immodest project, a multivolume work he planned to call *Universal History of the Characters of Letters and Languages of the Whole World.*

But at a certain point Kircher latched back on to the question of Egyptian letters, and to the notion held by Peiresc and others that Coptic might help crack the hieroglyphic code. In 1634, he persuaded Barberini to sponsor an additional translation project, despite maneuverings by Peiresc to try to secure another philologist for the job. The text, a Coptic–Arabic lexicon and grammar, had been brought back from Egypt some years before by a Roman gentleman named Pietro della Valle. He had begun his journey east in order to get over a broken heart; he then spent ten years traveling to places like Constantinople, Cairo, Jerusalem, Damascus, Ormuz, and Cyprus. In Baghdad he found a bride, who died shortly after they were married, and he traveled with her corpse, many live specimens of Persian cats, and many other souvenirs, for another five years before returning to Rome.

It was subsequently reported that Kircher was preparing a manuscript for publication. "Father Atanase Kircher is having his Egyptian language dictionary printed," one learned Frenchman wrote to Mersenne, "which will be like a precursor to his interpretation of the obelisks." And in 1636 he did publish such a precursor, the long Latin title of which translates as *The Coptic, or Egyptian Forerunner,*

*in Which Both the Origin, Age, Vicissitude, and Inflection of the Coptic
or Egyptian, Once Pharaonic, Language, and the Restoration of Hieroglyphic Literature Are Exhibited by a New and Unaccustomed Method.*
Although it had a main argument—that Coptic, which descended
from the language of the ancient Egyptians, had many helpful simi-
larities with that language system—it was otherwise a hodgepodge.
It quoted Barachias Nephi, but it couldn't remotely be said to contain the translation the cardinal had assigned him. It contained a
sample translation of the della Valle grammar, but not the whole
thing. It contained a partial, Hermes Trismegistus–sounding interpretation of the ostensibly ancient Bembine Tablet and of other
artifacts that didn't seem to relate to Coptic, or to Egyptian. One of
these was a strange inscription that had been found at the foot of
Mount Horeb in the Sinai Desert. Kircher said it was written in an
antique form of Chaldean used by Hebrew kings. According to his
wishful interpretation, these Hebrew kings somehow foretold that
"God will make a virgin" and "bring forth a son."

It's unclear what Cardinal Barberini thought of this book, but it
brought Kircher a certain amount of respect and Peiresc a certain
amount of relief (before he died the following year). In the view of
the papal censor, the text contained "many arguments brought forth
ingeniously from the hidden sanctuaries of holy antiquity and the
mysteries of the Egyptians." The author displayed "genuine knowledge of many languages as well as erudition in secret exotic matters."
It was, in short, "a worthy beginning from which we may anticipate
what will follow."

Mainly, in the words of a twentieth-century Egyptologist, the
Coptic Forerunner "unabashedly proclaimed that the decipherment of
the hieroglyphs was at hand." Kircher had given readers only a small

taste of what was to come, and toward the end of the text he announced the title of his big book on the subject: *Egyptian Oedipus.* Before killing his father and marrying his mother, Oedipus had solved the riddle of the sphinx, and now, as Kircher declared, he was like Oedipus—on the verge of solving one of the oldest riddles of all time.

8

Habitation of Hell

In 1637, not long after the publication of the *Coptic Forerunner*, a young German prince, the Landgrave Frederick of Hesse-Darmstadt, converted to Catholicism and made the journey to Rome. Among his other honors, he was to receive the Grand Cross from the Knights of Malta, the religious military order that traced its roots back five hundred years to the First Crusade. A German-speaking priest was needed to accompany him as confessor, and so, as Kircher recalled, "it happened that, a few years after my arrival at Rome, I was sent away to Malta with the Prince Landgrave."

Malta is really an archipelago—two big islands, Malta and Gozo, and many other small ones—about sixty miles off the southern tip of Sicily. It had been under the control of the Phoenicians, the Carthaginians, the Romans, the Byzantines, the Fatimids, the Normans, the Sicilians, and the Aragonese. After the Knights of St. John were ousted from the island of Rhodes in 1522, Charles V, the Holy Roman Emperor and king of Spain, ceded the property to the order for an annual symbolic rent payment of one live Maltese falcon.

Kircher must have wondered whether God would ever let him

stay in one place. But if he was initially upset about being dragged away with an immature prince to some rocks in the middle of the sea, his curiosity quickly kicked in. He made magnetic and astronomical readings, and studied geological formations. There were four-hundred-foot cliffs, natural arches, and a place where the tides had carved human-looking shapes into the earth. He explored Malta's megalithic temples, catacombs, and grottoes, and was especially fascinated by its inland seas and underground passageways: how far down did they go? Kircher observed the habits of Malta's troglodyte population, and on Gozo he climbed to one particular cave, high over Ramla Bay; some refer to it as the place where Odysseus spent seven years as the love slave of Calypso.

He also befriended a little man named Fabio Chigi, with whom he would exchange letters for the next two decades. Born into a rich and high-ranking family in Siena, Chigi was currently Malta's papal representative and chief inquisitor, and, though it was hard to imagine at the time, he would one day become pope. A frequent deviser of new ciphers and secret codes for use in official correspondence, he enjoyed writing poetry and studying arcane Etruscan inscriptions. Kircher and Chigi shared an interest in esoteric arts, and they must have discussed the so-called combinatory method of the thirteenth-century Majorcan theologian Ramon Llull, since it seems to have been the basis for a device that Kircher built and bestowed upon the knights.

Llull's system, set down in his tome *Ars Magna* (*The Great Art*), was itself derived from an Arabic technique for consulting astrological tables and from the practice of Kabbalah—in which, for example, the letters of the Hebrew alphabet are combined, contemplated, recombined, and contemplated again. His goal was to create a perfect language by which the entire universe, and the righteousness and inherent logic of Christianity, could be communicated to anyone. The tools included what are sometimes called volvelles, or wheel charts, concentric discs labeled with concepts or categories; the discs could be turned and aligned with one another to create a seemingly endless supply of new syllogistic statements about such things as God, goodness, greatness, and the nature of creation.

What is known about Kircher's device comes from an instructional guide he wrote titled *Specula Melitensis* (*Maltese Observatory*), which, as one historian says, mostly conveyed Kircher's "enthusiastic capacity for fatiguing detail." The apparatus had "the form and figure of an observatory," or watchtower, hence its name, and it evidently employed the Llullian discs. Beyond that, it's hard to say precisely what this instrument looked like or how it worked. A "universal chronoscope" was on "the first cubical side." A "cosmographic mirror" was on the second. A "physico-mathematical mirror" was on the third, and the fourth cubical side was used for "medical-mathematical" purposes. The top of the structure was a pyramid. In all, the device had one hundred twenty-five functions. Among other things, it could be used to determine:

- the "amount of dusk"
- the "flux and reflux of the seas"
- the astrological houses of the planets
- the signs of disease and "simple medicines for healing"
- the best times to go fishing and to give birth

"Receive it, Generous Knights," Kircher implored his Maltese hosts in the instruction book, "and receive it with favorable minds and eyes!" But there's no way to know if the device was well received, or ever used.

DESPITE ALL THIS ACTIVITY, after about a year on Malta in the service of the reportedly callow Frederick, and away from his work on hieroglyphics, Kircher started to worry that his potential would go unfulfilled. New thoughts about traveling to the source of ancient texts began to develop. As he'd done almost a decade before when feeling trapped in Mainz, Kircher wrote to the superior general of the Jesuits asking for a reassignment to Egypt or to the Holy Land. This request was denied, too, but another priest was found to replace him as Frederick's confessor, and Kircher was allowed to travel back to Rome. Once he was free of the prince and his retinue, however, he lost his sense of urgency, and he took his time getting there.

Kircher lingered in Sicily for a long while. Maybe this had something to do with wanting to reunite with Kaspar Schott, his younger friend from Würzburg. After fleeing from the Swedish army, Schott spent a few years in Tournai, and had ended up at the Jesuit college in Palermo. But there's no doubt that Kircher was fascinated by the natural formations and phenomena of the place. "I found such a Theater of Nature, displaying herself in such a wonderful variety of things, as I had with so many desires wished for," he wrote later. "Whatever thing occurs in the whole body of the Earth that is wonderful, rare, unusual, and worthy of admiration, I found contracted here."

He was especially intent on exploring Sicily's outcroppings, cliffs, and volcanoes. And he wanted to look into stories about a type of fish that lived in the Strait of Messina, the body of water that flows between Sicily and Calabria on the Italian mainland. The fish was supposed to be susceptible to a certain kind of song "by which," Kircher wrote, "mariners are wont to allure it to follow their vessels." But those plans had to be put aside because of the earthquakes that devastated much of Calabria in the spring of 1638. Kircher's account of events was published almost thirty years later and paid homage to Lucretius, Virgil, Lucan, and Dante.

Kircher recalled that the earthquake began on March 27, as he and some others crossed the strait by boat. The sea was "raging beyond what is usual" and—somewhere between mythical Scylla and Charybdis—began "stirring up huge whirlpools." The island volcano of Stromboli was "throwing up huge billows of smoke," and there was "a certain subterranean lowing, if you will, which we were reckoning to be the cracking of the earth and which seemed to conspire with the odor of sulfur to insinuate the complete, fatal and funereal destruction of Calabria and Sicily."

Despite a "cracking racket" and a "noisome odor," and the fact that the "sea itself was boiling," Kircher and his party made it across to Tropea on the mainland side, where there was, as there always seemed to be, a Jesuit college. They were there only a short time when "to such a degree did the violent and fearsome movement of the earth raise up a subterranean racket and din, similar to chariots driven at top speed, that the college along with the town at the foot of the mountain seemed to totter in the balance."

The earth "leapt up from below with so forceful a motion that I, no longer able to stand on my feet, was laid low, suddenly dashed down with face flat on the ground.

"O how at this point of crisis did the joys of the earth seem void of understanding!" Kircher wrote. "How at the bat of an eye did all honor, dignity, power and wisdom seem nothing other than smoke, a bubble, or straw snatched up by the wind!"

Amid "the crashing of the falling tiles and the creaking of the gaping walls," Kircher prepared to hand over his soul, even sensed it being "loosened from its corporeal fetters in order to take hold of the enjoyment of an unsullied existence." It was only after some time that he realized he wasn't hurt, and "resolved to venture for safety," running as fast as he could back toward the water again. "I reached the shore, but almost terrified out of my reason," he remembered. "I did not search long here, till I found the boat in which I had landed, and my companions also, whose terrors were even greater than mine."

On the next day, after experiencing "the intolerable frenzy of the earth" again in the form of an aftershock, Kircher's group sailed farther up the Calabrian coast. Stromboli was "raging in an uncustomary manner," and "the entire island seemed full of fires." They were coming ashore near another town when a groan from within the earth grew louder and louder. Finally it "struck the ground with such noise and indignation" that all were knocked off their feet, and the town they had been approaching was enveloped in a giant cloud of dust and debris. "After the cloud had dissipated little by little, we sought the town, but we did not find it," Kircher wrote. "A most fetid lake had been born in its place."

Through subsequent days of walking, they "came upon nothing but cadavers of cities and the horrific ruins of castles," he remembered. "Considering the men straggling through open fields as if extinguished for their fear, you would have said that at that very moment the day of final judgment was looming."

KIRCHER'S FIRSTHAND EXPERIENCE of this earthquake, which killed something like ten thousand people, might have put him off his investigations into the "miracles of subterraneous nature." But these horrible occurrences had also presented him with an opportunity for study. He was beginning to develop theories about the structure and workings of things below the surface of the earth and was eager to test them. "After having diligently searched out the incredible power of Nature working in subterraneous burrows and passages," he wrote, "I had a great desire to know whether Vesuvius also had not some secret commerce and correspondence with Stromboli and Aetna."

There was only one way, in his view, to find out. Vesuvius at that time was merely smoking. But its first major eruption in centuries had occurred fairly recently, in 1631. Kircher hired "an honest country-man, for a true and skillful companion," and the two began hiking their way up to the forty-two-hundred-foot summit at midnight. (Perhaps the reason for leaving at that hour was to be able to see in the dark anything that might be molten. Or maybe the idea was to allow for a full day of exploration once they got there.) The way was "difficult, rough, uneven, and steep."

When they finally reached the top, Kircher looked down into the crater. "I thought I beheld the habitation of Hell," he wrote, "wherein nothing seemed to be much wanting besides the horrid fantasms and apparitions of Devils." He heard "horrible bellowings and roarings" and there was "an unexpressible stink." The smoke

Mount Vesuvius, from Kircher's Underground World

and fire and stench "continually belch'd forth out of eleven several places, and made me in like manner belch, and as it were, vomit back again, at it."

When the morning light came, Kircher recalled, "I chose a safe and secure place to set my feet sure upon, which was a huge Rock, of a plain surface, to which there lay open an avenue, by a descent of the mountain very far. . . . And so I went down unto it."

The inside of the volcano was "all up and down everywhere, cragged and broken." But there was no gradual decline; the volcano's chamber was "made hollow directly and straight." The bottom was "boiling with an everlasting gushing forth, and streamings of smoke and flames, and employed in decocting Sulphur, Bitumen and the melting and burning of other kinds of Minerals."

Because the vapors and gases "know not how to be contained" within the molten matter, they did so "scatter the burden that lay upon them, with such great force and violence, accompanied with horrible cracklings and noises, that the mountain seemed to be tossed with an earthquake or trembling." Those spewings caused "the softer parts of the Mountain," made of, Kircher suggests, ashes, cinders, rains, and "the refuse of minerals," to be shaken to pieces and loosened; they fell "like Hills, into the bottom of the Hellish gulph." And *that* made the kind of sound that even "the stoutest and most undaunted heart would scarce venture to suffer."

Within this hollow mountain Kircher began to imagine what it might be like even deeper within the earth, and how the mountains and fires and rivers and oceans might somehow all be connected, as if they belonged to a kind of organism, or "geocosm," to use a word he would later coin.

9

The Magnet

Kircher arrived back in Rome with manuscripts and souvenirs for Cardinal Barberini, only to find that the cardinal wanted nothing more to do with him. As Kircher later spun the story in correspondence, it was Barberini who "delayed" him in Rome in the first place and charged him with producing "a hitherto unattempted work," and now he had "not only abandoned all memory of me, but also abandoned any concern for all the studies and books that he had promised."

Barberini's growing debts certainly had something to do with the decision to cut off Kircher's funding. It didn't help that Kircher planned to use the cardinal's money for an ostentatious work with a self-aggrandizing title. But there was also the question of Kircher's competence. He admitted that some people ascribed Barberini's decision "to my powerlessness and incapability and insufficiency." Humiliated and reduced to very dark "states of spirit," Kircher may have wondered whether the entire trip to Malta had been Barberini's way of getting rid of him.

But Kircher had a way of falling up. He found himself relegated to accepting one of the most prestigious scholarly positions in the Society of Jesus or anywhere else: the chair of mathematics at the Collegio Romano, the position formerly held by Christopher Clavius (deviser of the Gregorian calendar). Kircher moved into the quarters Clavius once occupied, a series of rooms off the college's second-floor colonnade. His cubiculum, as it was called, was much bigger than the professional cubicle of the twenty-first century. It was the place where he worked and slept: his bedchamber, laboratory, library, and workshop. When he first occupied the space, it already contained astrolabes, sextants, telescopes, clocks, and certain curiosities. There was a trick lantern, for example, that worked whether filled with oil or with water. Kircher added his own instruments, books, and manuscripts, as well as limestone stalactites, ostrich eggs, samples of pumice stone, and other things he'd collected on his trip south.

As long as circumstances "held me in Rome," Kircher wrote in his autobiography, "I decided that I ought to attain a reward for my trouble." That is to say that if his spectacular work on the hieroglyphs had to wait, he would in the meantime do some spectacular things in the field of mathematics.

Although distraught over Jesuit involvement in Galileo's prosecution, certain intellectuals in Europe had recently decided that it was the Jesuits who might be able to solve one of the biggest scientific problems of the era: how to figure out longitude at sea. Latitude, how far north or south you were, could be determined by the position of the sun at noon, the duration of daylight, or the height of the North Star above the horizon. But determining degrees of longitude, how far east or west you were, was extremely problematic. In

1598 the Spanish king had offered a major monetary prize—today's equivalent of about half a million dollars, plus almost two hundred thousand a year for life—to the person who could discover a reliable way of doing it. One possibility hinged on better information about magnetic variation, the degree to which the compass needle differs, according to geographical location, from the true north of the North Star.

About a year after Kircher returned from Malta, the intellectual instigator Marin Mersenne wrote from Paris to urge him to coordinate an effort from Rome: the Jesuits needed to get "someone in each college of the entire Society, by whatever means possible, to note accurately the variation of the magnet and the height of the pole star," meaning the variation at each latitude. "If this task were completed and if the authority of the supreme pontiff should lend itself to it," he wrote, "the result would be that at some time under the happy auspices of Urban VIII we would know the magnetic variation of the whole world, the altitudes of the pole, and the longitudes so long sought after."

Over the next two or three years, Kircher left virtually no aspect of magnetism untouched or unconsidered, and took it upon himself to head the collaborative enterprise Mersenne had described—one of the first attempts to collect what would now be called scientific data on a worldwide basis. His letters of instruction were carried by post throughout Europe and by ship to colleges and missions in places like Goa, Guadeloupe, Macao, Manila, São Paulo, and St. Augustine. Not every venue had the right instruments or the right expertise; one Jesuit in Lithuania who sent in variation readings worked as the cook in his college. In the course of directing this project, Kircher began to establish himself as a central contact and clearinghouse for Jesuit findings and reports on all manner of natural philosophy subjects.

After a while, when complicated inconsistencies in magnetic readings began to fade hopes for using them to determine longitude, Kircher became less interested in that particular project than in producing an impressive, elaborate, all-encompassing book on magnetism, one that would, as he put it, "rattle my adversaries' distrust of my work." He decided that if Cardinal Barberini didn't want to support him, then he'd just have to go to the Holy Roman Emperor instead. Through a Jesuit in the court of Vienna, he was able to secure the help of Ferdinand III, who agreed to fund the publication of a volume with many engravings and printed with special typefaces. A student of languages and a composer of music, Ferdinand was then married to the first of two first cousins who would bear him children, Maria Anna of Spain.

When it was published in 1641, Kircher's finished book—*The Magnet, or the Art of Magnetics, in Three Parts, in which the Universal Nature of the Magnet as well as Its Use in All Arts and Sciences Is Explained by a New Method: In Addition, Here Are Revealed through All Kinds of Physical, Medical, Chemical, and Mathematical Experiments, Many Hitherto Unknown Secrets of Nature from the Powers and Prodigious Effects of Magnetic as well as Other Concealed Motions of Nature in the Elements, Stones, Plants, Animals, and Elucescent Things*—came in at 916 pages.

In addition to compiling global magnetic data, describing practical magnetic aids to cartography, coining the word *electromagnetism*, discussing the magnetic quality of romantic love, and many other things, *The Magnet* (*Magnes* in Latin) took on the heliocentrists.

Both Kepler and Galileo had turned to magnetic principles to make their arguments for a sun-centered universe, but no one had directly refuted them on behalf of the pope, who, though he was once himself inclined toward the Copernican view, needed to strengthen the case for the decision against Galileo. Kircher, who once hinted to Peiresc that *he* was a Copernican, understood the position he was supposed to take. "We must always maintain that the white I see, I shall believe to be black," Ignatius had written, "if the hierarchical Church so stipulates."

Gilbert had said that the Earth was a giant magnet, spun and pulled around the sun by its magnetic, spiritual, cosmic, animate rays. Kepler had agreed. Kircher declared this notion *"absurda, indigna, et intolerabilis."* Earth wasn't a magnet, he argued, it was just, in certain ways, more or less, *magnetic.* (In this he was, more or less, correct.) Kircher calculated that if the little terrella that Gilbert had experimented with could attract, say, one pound, a magnet the size of the earth could attract more than three octillion pounds. The figure he gave was 3,073,631,468,480,000,000,000,000,000.

"Woe to all iron implements," he wrote, "woe to all shod horses and mules, woe to all soldiers in armor, woe to Gilbert's kitchen utensils."

Kircher argued that to the extent that there was magnetic force at work in the earth, it actually helped to *hold* it in place, right at the center of the universe, as the planets and the sun moved around it. Even if the sun did rotate and emit a magnetic effluvium to the planets, it wouldn't put them into perpetual orbit; a spinning magnet wouldn't put a magnet into orbit around itself. In general, Kircher declared, the comparisons to the magnet didn't hold up. And if Kepler "wished to philosophize prudently and consistently," he concluded, "he ought not to have gone beyond the limits of his analogy, lest he incur infinite contradictions and inextricable difficulties, which in truth he did."

Kircher was actually a great proponent of the magnetic power of the sun, and so incurred some contradictions and difficulties of his own. *The Magnet* was meant to serve not only as a weighty testament to its author's own world-class intellectual capacity but as a major argument for the pervasiveness of the magnetic principle, for the natural presence of the unseen. As Kircher saw it, the magnet was "that prodigal of nature, the true ape of the skies, the Idea of the universe in which whole new worlds are hidden, a divining rod and key to the unexhausted and undiscovered riches of the world." In many areas, Kircher shared or lifted the views of Mersenne's nemesis, the famous Dr. Fludd. Fludd believed the universe itself possessed a kind of magnetic or sexual energy, operating through attraction and repulsion, sympathy and antipathy, on a spiritual and physical level. Magnetic attraction was a "coition or union" between bodies caused by the similarity of their nature. The lodestone, Fludd wrote, "sucketh and attracteth from his center the body of Iron unto it, drawing forth of it his formal beams, as it were his spiritual food." In a sense, for Fludd and Kircher, it wasn't the earth that was a magnet; the magnet was God, and God worked in magnetic ways.

KIRCHER AGREED with Gilbert that magnetic polarity made young seedlings send their shoots up and their roots down. He likened the emanation of a flower's invisible fragrance to the emanation of a lodestone's invisible magnetic rays; that is, they were both somehow living spiritual emissions. And he believed, as he had previously attempted to show with his sunflower-seed clock, that trees and plants grew toward the sun (and in some cases toward the moonlight) by virtue of invisible magnetic attraction. He'd exchanged letters with Jesuits in Africa and Asia about exotic species that might manifest this principle. At the Collegio Romano he'd observed acacia, whose leaves opened up with the sun and closed at night. But of course it was the sunflower, brought to Europe from the Americas (the Incas venerated it as part of their sun worship), that figured most prominently in his thinking. Kircher admitted that his attempts to make workable clocks from sunflowers and sunflower seeds had been somewhat beset by complications. But he was apparently so convinced of the truth of the principle that a new fib was in order to help make his case. He wrote that he'd procured a "kind of material"— the root or seed of a heliotrope of some type—from an Arab merchant on the docks of Marseille. This material was more sensitive to the sun, he claimed, and worked much better than a regular sunflower seed did to drive a clock.

As Kircher described it, herbal and mineral remedies operated by magnetic action; ingested medicine "pulls what is similar to its own nature and ultimately draws and purges it." Antidotes worked in the same way. The best treatment for snakebite: eating the meat of a snake, preferably the one that bit you. (Kircher claimed that he'd seen this work back in Germany, on a traveling salesman from Erfurt.) Even the poison in snake venom itself arrived there in the first place through magnetic means; it was taken up through snakes' bellies as they slithered around on the ground. By its (Aristotelian) nature the snake had an appetite for the toxic mineral and vegetable excretions in the soil that were "putrid, contagious and noxious to men." The natural job of all venomous creatures was to siphon up poisonous vapors and emissions, leaving the earth safe for human beings.

Spiders drew in their venom from the very air. In a section of *The Magnet* that Kircher said was worth the price of the whole book, he argued that magnetism was at work in the well-known cure for the bite of a certain spider found around the southern Italian town of Taranto. This "tarantula," as it is called, has almost nothing in common with the much larger, hairy, and dangerous North American spider that was later given the same name. In fact, the Italian tarantula is now known to be basically harmless. But in Kircher's time, every summer, people who claimed they were bitten by the tarantula exhibited an array of troubling symptoms: delusions (imagining themselves as expert swordsmen, for example, or as ducks or fish), listlessness, jumpiness, twitchiness, giddiness, lethargy, unusual and excessive thirst for wine. Afflicted women ran around exposing themselves. Men experienced unrelenting erections. They could be cured only by certain kinds of up-tempo songs, "tarantellas," to which they responded involuntarily in the form of a frenetic dance. The playing and dancing went on for hours. The cure could take anywhere from three to eight days. During this time villagers who had been bitten in *previous* years often experienced a recurrence of the disease from hearing the music, and began dancing as well.

In a way that exemplifies the pre-modern approach to scientific matters, the existence of the disease and the efficacy of the treatment were not really in doubt. The question was not whether the music worked—or whether the bite caused such symptoms in the first place, or whether people might actually *want* to get "bitten" so that they could indulge in the cure. The only question was *how* the music worked, and many intellectuals speculated about the answer, from a

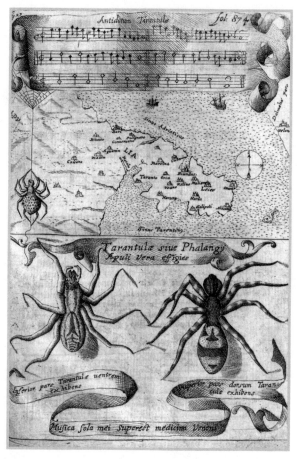

The tarantula and its musical antidote, from The Magnet

great distance away, without anything like firsthand observation or methodical reports. One natural philosopher from Naples believed that the spider's venom increased the temperature of the "spirits" in the bloodstream to the level at which they would be inclined to dance and hop like a spider; the music attracted the spirits out of the body. Even Gassendi, the atomist, was happy to provide an explanation, one that was consistent with his more modern, physical philosophy: the music pushed motion onto the blood, which pushed motion onto the muscles and nerves, resulting in the dancing; the impact eventually broke down the particles of poison into ineffectual bits. Kircher, the new champion of the magnetical philosophy, held that it was the music itself that did most of the work, and that it did so magnetically—drawing the poisonous humor from the deep fibrous recesses of the body so that the sweat from the dancing could carry it out. As Kircher understood it, the same music didn't work for everybody; sanguine types, for example, were sympathetic to the soft sound of the zither, but the way to draw the poison out of cold, phlegmatic sorts was to agitate the venom with drumbeats and cymbals.

Despite what seems from a modern perspective like major susceptibility to nonsense on Kircher's part, he was skeptical about many things that Fludd, for example, was not. Fludd was a great champion of the weapon salve that was supposed to heal a wound from any distance. But Kircher didn't believe it. No natural force was without physical limitations, he argued, and the natural force of magnetic attraction was no exception. He concluded that if cures actually resulted, they were miracles performed by God or by angels, the result of some demonic art or black magic, or they were natural but unrelated to the salve.

Kircher refuted the notion that consumption could be transferred to a dog or pig by feeding the animal an egg boiled in the blood of the patient. And he dismissed commonly held beliefs about something called the vegetable lamb plant of Tartary. By many accounts, this Central Asian plant grew actual sheep as its fruit, with a soft coat outside and bloody meat inside. The grasses around the plant always appeared to have been eaten, though usually the lamb fruit (or fruit lamb?) itself was too high off the ground to have reached the grasses, and some people theorized that magnetic forces helped draw them up within reach. Kircher had never seen one of these plants in person but conjectured that the lamb wasn't an actual lamb—the fruit just looked like a lamb, the way the fruit of a plant in distant California was said to look like a dragon. In his opinion the grass was merely stunted by the lamb plant's everyday suction of healthy substances from the soil.

IT WAS IN PART because of Kircher's skepticism on matters such as the lamb plant of Tartary, his willingness to separate lore from the literal truth, that *The Magnet* "earned not insignificant applause," as he put it, from many intellectual quarters. The book was so well received that a second, enlarged edition was printed two years later. Among those who were adopting a more empirical approach to knowledge, however, it was valued more for the entertainment than the information it provided. After *The Magnet* appeared in printed form in 1641, a young Italian intellectual sent a report about it to Galileo, who was now old and blind. It was "a very large volume on the magnet," he wrote, "a volume enriched with an abundance of beautiful copperplate engravings. You will see astrolabes, clocks, wind scopes, with a flourish of extremely outlandish names. Among other things there are . . . inscriptions in Latin, Greek, Arabic, Hebrew, and other languages. One delightful thing is the musical score he claims is the antidote to the tarantula's venom." In short, the Italian and his colleagues "had a good laugh."

Others were curious to see what it contained. "I am approaching the point where I have to deal with the lodestone," René Descartes wrote to the polymath Constantijn Huygens in 1643. "If you think that the big book you have on the subject, of which I don't know the name, could be useful to me, and you would not mind sending it to me, I would be much obliged, and will be for the rest of my life."

Two days later, Huygens replied, sending him "the Magnet by Kircherus, in which you will find more grins than real substance, which is ordinary coming from the Jesuits." He also "begrudgingly" sent along books by Gassendi, asking Descartes to return them quickly because he would not be able to learn anything from them: "The nonsense of fools takes as much time to read as the good things of the learned."

Many years before, in a stove-heated room somewhere in Germany, Descartes had had a vision for an entirely new method for acquiring knowledge: it could be built only on what was without doubt and what was mathematically certain. What Descartes did, among other things, was to cut off consideration of immaterial influence on the material world. He didn't take God off the table—in fact he "proved" God's existence, as well as, very famously, his own ("I think, therefore I am")—but he limited explanations of natural phenomena to physical, mechanical causes, cutting off the realm of the body and the physical world from the realm of the mind and soul and spiritual world. And yet Descartes's own explanation for magnetic attraction and polarity was utterly speculative and highly uncertain; it involved the constant flow of particles that he imagined to be threaded like screws. The threading was such that the particles could enter one pole of the magnet only through entry holes that were likewise threaded and exit only through threaded exit holes at the other end.

A week after receiving the books, having spent some time "flip-ping through them," Descartes sent them back. "I believe that I have seen everything that they contain, even if I have hardly read anything but the titles and margins. The Jesuit is quite boastful; he is more of a charlatan than a scholar."

Descartes went on to comment about the material that Kircher said he'd gotten from the Arab merchant in Marseille and that was supposed to turn toward the sun day and night. "If it were true, it would be interesting, but he does not explain at all what the material is," he wrote. He remembered that "Father Mersenne wrote to me in the past, about eight years ago, telling me that it was from the sun-flower seed, which I do not believe, unless the seed is more powerful in Arabia than it is in this country."

It was absurd, and yet not impossibly absurd. Descartes tried it himself, he wrote, "but it did not work."

10

An Innumerable Multitude
of Catoptric Cats

Pope Urban VIII died in the summer of 1644, and was succeeded by Pope Alexander VI's great-great-great-grandson.
(Alexander VI was evidently not a great believer in celibacy.)
Choosing the name Innocent X, Giovanni Battista Pamphilj took
quick legal action against the Barberinis for embezzlement, forcing
Cardinal Barberini and other members of his family to flee to Paris
for a time.

During the fall of that year, a young English traveler watched the
procession of the new pope on its way to the Basilica of St. John Lateran. According to his report, first came "a guard of Switzers" and
the "avant-guard of horse carrying lances." Next came (putting his
words into line):

+ those who carried the robes of the Cardinals, two and two
+ then the Cardinal's mace-bearers
+ the *caudatari*, on mules

- the masters of their horse
- the Pope's barber, tailor, baker, gardener, and other domestic officers, all on horseback, in rich liveries
- the squires belonging to the Guard
- five men in rich liveries [leading] five noble Neapolitan horses, white as snow, covered to the ground, with trappings richly embroidered, which is a service paid by the King of Spain for the kingdoms of Naples and Sicily, pretended feudatories to the Pope
- three mules of exquisite beauty and price, trapped in crimson velvet
- three rich litters with mules, the litters empty
- the master of the horse alone, with his squires
- five trumpeters
- the *armerieri estra muros*
- the fiscal and consistorial advocates
- *capellani, camerieri de honore, cubiculari* and chamberlains, called *secreti*
- four other *camerieri*, with four caps of the dignity-pontifical, which were Cardinals' hats carried on staves
- four trumpets
- a number of noble Romans and gentlemen of quality, very rich, and followed by innumerable *staffieri* and pages
- the secretaries of the *chancellaria, abbreviatori-accoliti* in their long robes, and on mules
- *auditori di rota*
- the dean of the *roti* and master of the sacred palace, on mules, with grave, but rich foot-clothes, and in flat episcopal hats

+ more of the Roman and other nobility and courtiers, with diverse pages in most rich liveries on horseback
+ fourteen drums belonging to the Capitol
+ the marshals with their staves
+ the two syndics
+ the conservators of the city, in robes of crimson damask
+ the knight-confalionier and prior of the R. R., in velvet toques
+ six of his Holiness's mace-bearers
+ the captain, or governor, of the Castle of St. Angelo, upon a brave prancer
+ the governor of the city
+ on both sides of these two long ranks of Switzers
+ the masters of the ceremonies
+ the cross-bearer on horseback, with two priests at each hand on foot
+ pages, footmen, and guards, in abundance
+ the Pope himself, carried in a litter, or rather open chair, of crimson velvet, richly embroidered, and borne by two stately mules; as he went, he held up two fingers, blessing the multitude who were on their knees, or looking out of their windows and houses, with loud *vivas* and acclamations of felicity to their new Prince
+ the master of his chamber, cup-bearer, secretary, and physician
+ the Cardinal-Bishops, Cardinal-Priests, Cardinal-Deacons, Patriarchs, Archbishops and Bishops, all in their several and distinct habits, some in red, others in green flat hats with tassels, all on gallant mules richly trapped with velvet, and led by their servants in great state and multitudes
+ the apostolical *protonotari*, auditor, treasurer, and referendaries

- the trumpets of the rear-guard
- two pages of arms in helmets with feathers and carrying lances
- two captains
- the pontifical standard of the Church
- the two *alfieri*, or cornets, of the Pope's light horse, who all followed in armour and carrying lances
- innumerable rich coaches, litters, and people

The detail-oriented observer was John Evelyn, a former Oxford student who would rather have gardened than serve in the Royalist army. Although the English king Charles I was in the midst of a civil war against Oliver Cromwell's Parliamentarian forces, the twenty-four-year-old Evelyn had been given leave to travel, and he was now satisfying his curiosity about Rome. Later in life he wrote thirty books on such topics as forestry and the cultivation of fruit trees, the smoke and smog of London, copper engravings, English customs, French fashion, and ancient architecture. A couple of weeks prior to the papal procession, Evelyn sought out an introduction to Kircher, who, despite the smirks of the more astute new philosophers, had realized at least some of his ambition for fame as the author of *The Magnet*.

Not that Kircher had given up his sense of himself as the new Oedipus. He never really lost interest in anything, and almost ten years after starting it, he'd recently published the Latin translation of the Coptic lexicon and grammar that Pietro della Valle had brought back from his travels east. Roman dignitaries and foreign visitors such as Evelyn came to see him. Jesuit missionaries around the world sent him astronomical observations, reports of unusual phenomena, specimens of flowers, animals, and shells.

"Father Kircher . . . showed us many singular courtesies," Evelyn wrote in his diary, "leading us into their refectory, dispensatory, laboratory, gardens, and finally . . . through a hall hung round with pictures . . . into his own study, where, with Dutch patience, he showed us his perpetual motions, catoptrics, magnetical experiments, models, and a thousand other crotchets and devices."

Kircher's catoptric (mirrored) displays included his "catoptric theater," a cabinet with a reflective interior that appeared to multiply infinitely whatever was set inside it. It's possible that Evelyn had already seen one of these in Rome; there was one at the Borghese family villa and one at the palazzo of another prince. It was very amusing to watch uninitiated guests grab at the air where they thought they saw, say, stacks and stacks of gold coins. But Kircher's chest "far surpassed the competition," says historian Michael John Gorman. And he used it to fool all types of susceptible creatures.

"You will exhibit the most delightful trick," an assistant to Kircher later wrote, "if you impose one of these appearances on a live cat, as Fr. Kircher has done. While the cat sees himself to be surrounded by an innumerable multitude of catoptric cats . . . it can hardly be said how many capers will be exhibited in that theatre, while he sometimes tries to follow the other cats, sometimes to entice them with his tail, sometimes attempts a kiss, and indeed tries to break through the obstacles in every way with his claws so that he can be united with them."

Evelyn didn't mention any cats in his diary (or any capers), but by the early 1640s, Kircher's interest in mirrors, reflection, and optics had only intensified. There is evidence that during this period he began to entertain, or frighten, his visitors with a primitive type of magic lantern, projecting images of Satan and death onto the walls

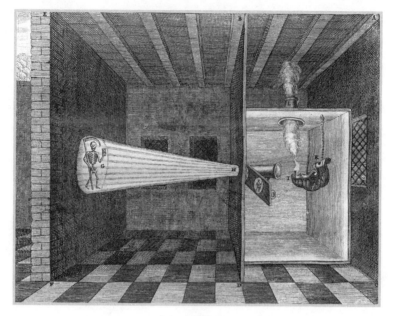

A magic lantern

of his darkened cubiculum. He is also known to have dissected the eyeballs of bulls in order to understand the way vision works. All these activities informed the production of his next book, *The Great Art of Light and Shadow.*

The title in Latin, *Ars Magna Lucis et Umbrae*, was intended as a play on words: "We say 'Magna' on account of a kind of hidden allusion to the magnet," Kircher wrote in his introductory pages, meaning that the title could also be read as "The Magnetic Art of Light and Shadow." He'd begun to see everything in more or less magnetic, almost binary, terms: attraction and repulsion, positive and negative, friendship and strife, light and dark. To Kircher the connection between light and magnetism seemed clear; it was the magnetic attraction of the sun's rays, after all, that made the sunflower turn toward it—and that, at least in theory, might make a sunflower seed turn toward it too.

Like *The Magnet*, this book was conceived as an encyclopedic work, an "encirclement" of the entirety of its subject, one that in almost a thousand pages, plus dozens of engravings, diagrams, maps, and illustrations, would provide readers with all they could possibly want to know about light, color, vision, and related matters. It also provided some impressive evidence of mystical erudition. The book's ten-part structure, as Kircher explained it, connected to the ten-stringed harmony of the Greek instrument the decachord. This in turn represented the well-ordered harmony of nature, and the Decalogue, or Ten Commandments, and the Pythagorean notion of the number ten as the number of the universe and perfection, as well as the Sefirot, the ten emanations of God, by which, according to Kabbalah, the universe was created. "For just as the wise men of the Hebrews claim a world built from ten rays of divinity," Kircher wrote in his preface, "so we completed ten separate themes or books, as it were, ten books in ten parted rays, the world of light and shadow, that is, our art."

Kircher wrote like Hermes Trismegistus himself when it came to describing God's purpose with respect to the sun: "He has established the Sun, I claim, as a kind of heart or soul, or a sort of intelligence and, I may also say, as the principal control and will of nature, in order that the world be governed by it, and that in order that the hidden sacraments of God's wisdom be revealed torn out of the chaos and abyss of darkness, and in order that, from the latter visible and material will, the majesty of the former invisible super-mundane will become known to mortals."

The book devoted hundreds of pages to sundials, and to Kircher's own theories about fireflies (they appear to have voluntary control of their flashes), chameleons (they stop changing color once they are dead), and phosphorescent jellyfish (they've simply been endowed with the ability to produce light so that they can see in the darkness of deep water). It provided readers with a calculation of the "thickness of the atmosphere"—forty-three thousand paces—undertaken by measuring "the refraction of sunlight in air," and offered advice to artists on perspective as well as "rules which must be followed in painting scenes and drawing pictures." There were musings on the color of angels, and on why the sky is blue: in order to provide a proper visual background for everything, it had to be "a kind of unequal mixture of light and dark," and, after all, blue is "a color by which the uninterrupted sight may contemplate that most agreeable space of the heavens."

Among other information on optical curiosities and devices, *The Great Art of Light and Shadow* significantly included one of the earliest published descriptions of a microscope. (Della Porta described magnifying lenses in *Natural Magic*, and Galileo wrote about the use of a telescope "adjusted" to see things close up. He'd made a gift of one to the experiment-minded members of the Accademia dei Lincei, or Academy of the Lynx-Eyed in Rome, who in turn had published descriptions and images of magnified bees. The Linceans, as they were called, may have been the source of Kircher's device.) Although Kircher's *smicroscopus* was not much more than a short tube with a magnifying lens, or lenses, inside it, he claimed to have seen "mites that suggested hairy bears" and minute organisms in cheese, vinegar, and milk. If the worm-like forms that can be seen through a microscope are "so tiny that they are beyond the reach of the senses," he wondered later, "how tiny can their little hearts be? How tiny must their little livers be, or their little stomachs, their cartilage and little nerves, their means of locomotion?"

But Kircher wasn't very accurate when it came to optical specifications for others to follow, and there's doubt about his technical expertise. In 1645, after he sent some kind of image-projection assembly to the emperor in Vienna, a Jesuit in the court wrote him twice for more precise instructions because he couldn't get it to work. Years later an Englishman reported that "an eminent man of optics" in Nuremberg "spoke bitterly to me against Father Kercherius, a Jesuit at Rome . . . saying that it had cost him above a thousand pounds to put his optic speculations in practice, but he found his principles false, and showed me a great basket of glasses of his failings."

Meanwhile, a Minim friar named Emmanuel Maignan complained in correspondence that Athanasius's work on catoptrics was a little too similar to his own. And after the censors approved the text of *Light and Shadow*, a satiric work called *Monarchy of the Solipsists* began circulating around Rome. Written by an obviously fictitious author, Lucius Cornelius Europeaus, the pamphlet made fun of the Jesuits of the Collegio Romano, and of Kircher, depicting him as an "Egyptian wanderer" who "broadcasts trifles about the Moon." The "Solipsists" consider theological questions, such as "whether the souls of the Gods have color," as well as philosophical questions, such as:

> *If a mouse urinates in the sea, is there a risk of shipwreck?*
> *Are mathematical points receptacles for spirits?*
> *Is a belch an exhalation of the soul?*

But overall, the response to Kircher's magnificent illustrated compendium was positive, and so was the way in which Jesuit authorities believed his prodigious intellectual output reflected on the order and on the Church. Pretty soon they relieved him of his teaching duties, not to keep students safe from the work of this Egyptian wanderer, but to free him to do more of it.

KIRCHER NOW SHIFTED his attention from the entirety of optics to the entirety of acoustics and sound. Or rather, as he later wrote, there was a "desire of joining to our work on Optics a work on Acoustic faculties with like variety and opulence of evidence." The reason was simple: "I discovered so great a mutual affinity between the two that I have concluded that light is nothing other than a certain consono-dissonance for the eyes, while sound . . . is a certain shadow-light for the ears."

In the course of producing his next massive tome, *Musurgia Universalis sive Ars Magna Consoni et Dissoni* (*Universal Music-making, or The Great Art of Consonance and Dissonance*), Kircher studied the ears and the vocal organs of humans, animals, and birds, and began cataloging the history of musical instruments and musical styles. He also joined with a craftsman named Matteo Marione to replace the old water-powered organ in the gardens of the Palazzo Quirinale. And he started experimenting with what he called "tone architecture"—"the reflection of sounds and the multiplication of the same."

Guided by the "great similitude of light and sound," he applied his (as it turned out, incorrect) understanding of reflected light to the reflection of sound. After a number of trials, Kircher developed "a conical tube, or, if you will, one projected into a spindle," that seemed to carry sound better than other designs. He subsequently installed a very large version of one of these tubes—"joined with iron sheets about twenty-two palms in length"—in the wall between his cubiculum on the second floor of the Collegio and the interior gate facing a garden downstairs.

"Should need arise for our porters to inform me of some matter," he explained, "whether concerning the arrival of some visitors or any other affair, instead of taking the trouble to reach my [quarters] through the various labyrinthine windings of the house, they would speak to me while standing within the security gate, as I lingered in the remote recess of my bed chamber, and, as if present, they reported distinctly and clearly to me whatever they might wish; and I, in turn, was responding to the matter with the same tone of voice through the mouthpiece of the tube."

It also came in handy as an eavesdropping device. "I daresay that no one in the garden was able to say anything in a voice above a whisper that I did not hear within my bed chamber," he recalled.

Among hundreds of other acoustical innovations and musical machines included in *Universal Music-making*, eventually published in 1650, the book described the glass harp (goblets arranged by the level of liquid within them, played by passing a moistened finger around the rims) and the Aeolian or wind harp (a box-shaped stringed instrument placed in a window and played by the breeze). One instrument that doesn't seem to have been included—at least no one can find it in the twelve hundred pages of Latin that make up the two-volume work—is the infamous cat piano with which Kircher has so long been associated. But Kaspar Schott, his friend and disciple from Würzburg, did publish an account of it. Schott wrote that a "distinguished and ingenious" person, who sounds like Kircher, constructed one of these pianos to dispel the melancholy of an unnamed prince, who sounds a bit like the Prince-Elector of Mainz. This person

> captured living cats, all of differing size and consequently of differing shrillness and depth of voice; these, in a certain chest constructed with effort devoted to this purpose, he enclosed in such a way that their tails, after they had been stretched through apertures, were fastened and led through to certain determined channels. Upon these he furnished keys constructed with most slender pricks in the place of mallets. . . . In proportion to their differing tonal magnitude he arranged the cats so that individual keys corresponded to their individual tails and he established the instrument in a place suitable for the relaxation of the Prince. When it was finally played, it produced the sort of harmony as the voices of cats are wont to supply. For when the keys had been depressed by the fingers of the Organist, since with their very pricks they punctured their tails, the cats, driven to a state of madness, thundering with piteous voice now deep, now shrill, were producing a harmony arranged from the voice of cats, which thing both moved men to laughter and was able even to drive the mice themselves to the fields.

Kircher was not really a musician; he played no instrument, feline or otherwise, but he was an opinionated aficionado of music at a time and place, baroque Rome, of extraordinary musical activity—not only the opera, but orchestration, and the trombone, for example,

The cat piano

were new. As a mathematician with training in the ancient study of harmonics, he had little patience for what he heard in the churches, chambers, and theaters by composers and singers who lacked rigor and grounding in musical intervals. Kircher conceded that "notable abuses and faults" were impossible to avoid "amid so great a throng of musicians." Still, he bemoaned the situation in which "such wretched compositions, prone to so many errors and defects, should appear every day, often even in the leading places." He also complained about hearing "the same twittering, the same cluckings, the same phrases everywhere until you feel sick and angry."

As a partial solution to the problem, Kircher offered readers a do-it-yourself system for writing music according to proper mathematical rules. "The mechanical production of music," he explained, "is nothing other than a certain closely defined method I have invented, by which anyone, even if he has no musical knowledge, may, by varied application of music-making tools, compose tunes." It was a project that he'd been working on since his time in Würzburg with Schott. Based on the combination techniques of Ramon Llull—the same techniques he'd employed to make his strange calculating machine on Malta—the procedure was actually limited in scope to four-part polyphonic settings for voice. Nevertheless, an almost endless number of compositions could result from combining hundreds of short musical phrases in various styles and pitches, which Kircher represented as values on numerical tables. Kircher even used the elements of the system to build a number of music-computing cabinets for presentation as gifts to certain patrons and dignitaries. These "musarithmetic arks," as he called them, were handsome wooden boxes containing many columns of long removable slats onto which the values had been imprinted; as the slats were pulled out, music was pieced together.

Most of the music theory and a lot of other material contained in *Universal Music-making* was actually lifted right from Marin Mersenne's large work on music. ("Father Kircher devoured my book *Harmonie Universelle* that I lent him here in four days," Mersenne wrote during a visit to Rome in 1645, "and says he is thrilled.") But for readers, that only added to the value of this encyclopedia, which served as the standard musical text in Europe well into the eighteenth century.

Kircher's original contributions to his own book are in fact well respected. In addition to creating new classifications of musical styles, Kircher was apparently the first to articulate the "doctrine of the affections" on which so much later baroque music operated. Kircher believed that music's great purpose was to echo and evoke human emotions or affective states, but that this purpose was best achieved through formal rhetorical technique and structure. A German translation published in 1662 made *Universal Music-making* especially important in the territories of the Holy Roman Empire, where, as the modern Italian composer Roman Vlad believes, it had "a more or less direct influence on Johann Sebastian Bach."

11

Four Rivers

Kircher's success with magnetism, optics, and music was "grounds for praise of God," but it also offered surprising "fodder for tribulation," as he put it, and may have even brought about some paranoia. (The real reason for installing an eavesdropping device in his bedchamber?) Those who had always been skeptical of him, he recalled, "attacked me anew with fresh accusations." The charge: He'd been concentrating on those other subjects because he wasn't getting anywhere with his most important work, "as if abandoning all hope of addressing Hieroglyphics on account of its impenetrable difficulty."

Kircher was now in his mid-forties, and it had been much more than a decade since he began trying to decipher the Egyptian system. In the meantime, his vision for *Egyptian Oedipus* had grown more and more ambitious, and more and more expensive to execute. It would have to be longer than he originally thought. Many exotic typefaces were required. More artists and engravers were needed to render the illustrations. Now, he said, God provided "an utterly marvelous manner" for him to resume his hieroglyphic work and to "elude the empty machinations" of his enemies. Events led not only to the publication of *Egyptian Oedipus* but to Kircher's collaboration with Gianlorenzo Bernini on what many people regard as his master-piece, the Fontana dei Quattro Fiumi (Fountain of the Four Rivers) in the Piazza Navona, one of the most well-known public squares in the world.

In 1647, Pope Innocent X decided, "for the immortality of his own name," as Kircher wrote, to re-erect an Egyptian obelisk in the center of the piazza, which occupies the oblong site of the old Circus Agonalis, a stadium where ancient Roman games were held. Innocent and other members of the Pamphilj family—including his powerful sister-in-law, Donna Olimpia, called the *papessa*, or "lady pope"—were in the process of combining their existing properties on the piazza into one great palazzo appropriate to their exalted sta-tion. The obelisk was meant to elevate the aesthetics and the stature of the entire square in advance of the Church's Jubilee Year celebra-tion in 1650.

"Since he had heard that I possessed skill in the Egyptian al-phabet and Hieroglyphics, and that I was called to Rome for that reason," Kircher recalled, Innocent "fetched me to himself." Those who met the pope tended not to forget it. Innocent was ugly. ("He was tall in stature, thin, choleric, splenetic, with a red face, bald in front with thick eyebrows bent above the nose," a Roman of the time reported. "His face was the most deformed ever born among men.")

"Father," Innocent said to Kircher, "we have decided to erect an obelisk, a stony mass of not small size. Yours will be the task of transforming it to life with your interpretation."

At the time, this obelisk still "lay thrown to the ground and bro-ken into five parts" outside town at the Circus of Maxentius, where it had been erected about twelve hundred years before. It was going to have to be made whole again. "And since from the corrosion of its letters the obelisk was greatly defective and several outlines of figures were lacking," Kircher explained, "His Holiness wished that it be restored to its unimpaired condition by putting upon me the task of filling in all the missing portions in accordance with my knowledge."

Kircher imagined his detractors whispering to each other: "Let us just see if he has experience in the knowledge of Hieroglyphics or if he can genuinely be of good use in addressing these figures." As if to prove to these skeptics just how much "experience in the knowledge of Hieroglyphics" he had, he quickly produced five hundred pages on the subject.

It was actually Francesco Borromini, Bernini's gloomy nemesis, who was expected to receive the commission for the obelisk's base. Innocent wanted a big fountain; he'd already put Borromini in charge of diverting water to the site from the Acqua Vergine, a still-functioning ancient Roman aqueduct, and he'd already given tacit approval to Borromini's concept: a fountain that would represent the four principal rivers of the four continents known at the time. Bernini was out of favor, not least because of his association with the Barberini family, but Innocent changed his mind about the commission after seeing a model of Bernini's stunning design. Today it's all the more stunning to consider, given that Kircher played an important role in its development. "I would even venture to say," writes one Bernini scholar, "that it was only through Kircher that Bernini . . . managed to displace Borromini."

Bernini stuck to the four-rivers idea, which was meant to represent the centrality of the Church and the flow of faith to all the corners of the globe, as if carried along on the waters of the Danube, the Nile, the Ganges, and the Río de la Plata. (The Americas were regarded as a single entity, and although the Portuguese had been navigating the Amazon for years, the Río de la Plata was more famous.) And for the details of animal and plant life along the rivers he'd never seen, he consulted Kircher.

Kircher had never seen them, either. (Were Bernini and others in Rome still under the impression that Kircher had actually been to the East? Maybe they were.) Nevertheless, he had an emerging reputation as an authoritative polymath. He was the figure in Rome with whom a growing number of other, some might say "actual," authorities on a range of subjects chose to correspond, and he was the Jesuit to whom missionaries most frequently sent reports, artifacts, and natural specimens. One of those specimens was the preserved body of what's now called a giant armadillo, a South American animal that looks even more unpleasant than its smaller cousin, with bony plating, scales, and large claws. In the well-known engraving of Kircher's museum, it's hanging from the ceiling. Kircher may not have been entirely clear on this animal's bearing and habitat, however, because it seems to have inspired the carved sea creature that stands upright in the water on the American side of the fountain, referred to as the "Tatu of the Indies" by a seventeenth-century chronicler of Bernini's life and work. (*Tatú* is Spanish for "armadillo.")

The effect of Bernini's fountain goes well beyond its fairly straightforward imagery and symbolism (there's a dove on the very top of the obelisk, for example, that represents the pope). The rocky travertine base is made to look like a crush of tectonic forces, a mountain in the making. Water shoots from cracks and crevices all around it, and it's carved out in such a way that you can see through it on all four sides.

Each main chunk of the base supports one of four giant marble river gods, but the interior of the form on which the fifty-five-foot obelisk rests contains mostly a lot of water and empty space. As the same seventeenth-century writer put it, "One marvels not a little to see the immense mass of the obelisk erected on a rock so hollowed out and divided and observe how—speaking in artistic terms—it seems to stand upon a void."

The concept is evidently based on conclusions that Kircher had reached since exploring the caves, underground seas, and passageways of Malta and Sicily, not to mention the deep crater of Mount Vesuvius. As he later explained, "The whole Earth is not solid but everywhere gaping, and hollowed with empty rooms and spaces, and hidden burrows." Deep down, it held many oceans and fires, interconnected by passages that reached all the way to its core, and there were many entrances and exits in the ground and the floor of the ocean. In the case of the latter, enormous quantities of water passed back and forth through them. "The sea," he wrote, "by the winds and pressure of the air or the motion of the estuating tides, ejaculates, and casts the waters through subterraneous or underground burrows into the highest waterhouses of the mountains." In other words, Kircher believed that mountains were hollow and served as great reservoirs. Water pushed its way out through the sides of the mountains like the water of Bernini's fountain, flowing down the slopes as rivers and streams, and completing the cycle by emptying into the seas.

But the fountain evokes an even larger and more primal natural process, as if capturing the moment when an animating force or a shock of the divine permeates the material world. Bernini may have known that obelisks, with their long, tapering shapes, were said to represent rays of sunlight. And Kircher would have been pleased to inform him of the magnetic manner in which he believed the rays of the sun gave life, the way in which sunlight acted as "the lodestone of heaven, drawing all to it." Light "passes through everything," he wrote in *The Great Art of Light and Shadow*, and "by so passing through, it shapes and forms everything; it supports, collects, unites, separates everything. All things which either exist or are illuminated or grow warm, or live, or are begotten, or freed, or grow greater, or are completed or are moved, it converts to itself." Not by chance, Kircher's reading of the hieroglyphics on the obelisk revealed that the ancient Egyptians, inscribers of sacred wisdom, basically agreed: In brief, his translation describes the emanations of a "Solar Genius" and the lower entities that are "drawn," "fructified," and "enriched" by the diffusion of its energy.

The important thing is that Bernini wouldn't have been totally bemused by all this. Kircher's animistic, Neoplatonic, and catholic views were also still somehow Catholic, and they were characteristically baroque. According to art historian Simon Schama, Bernini "was forever inventing new ways in which the unification of matter and spirit, body and soul, could be visualized and physically experienced." While "it is difficult to trace the exact degree of closeness between the sculptor and the Egyptologist, something like this belief—the revelation of divinely ordained unities, tying together the different elements of living creation—is surely the controlling concept behind Bernini's immense creation."

WORK ON THE FOUNTAIN STARTED—with some pieces of the obelisk dragged by water buffalo, and others moved with winches, canvases, and horses—as Rome began to suffer from a grain short-

age. When taxes were levied to help pay for the project, people became angry. Rhymes of protest began to appear overnight on the
blocks of travertine and the chunks of obelisk lying in the piazza: the
people didn't want *fontane* (fountains), they wanted *pane* (bread).
Jesuits began giving out bread, soup, and wine to the poor, along
with tickets that could be turned in for alms at a later date. When
that day came, so many people showed up that "a terrible thing happened," a Roman man reported. "Some of the poor died, suffocated
by the crowd; a pregnant woman fell unconscious, others broke their
legs, and in short many were injured."

General hunger notwithstanding, the fountain kept going up, and
the Pamphilj pope paid for the publication of Kircher's lavish *Pamphilian Obelisk*, a book that serves for all intents and purposes as the
first volume of his *Egyptian Oedipus*. Kircher used it as a kind of sales
tool. As he told it, when Emperor Ferdinand III in Vienna got a look
at the handsome tome that Kircher had dedicated to Pope Innocent,
"he sent to me most eloquent letters by which he kindly impelled me
to take up anew my work on *Oedipus Aegypticus*." Through an inter-
mediary, Kircher explained to Ferdinand that because of all the special typefaces and necessarily elaborate engravings, the book "would
not be able to be produced for less than three thousand scudi," almost $200,000 in twenty-first-century money. Ferdinand agreed to
release the "altogether worthy amplitude of his munificence" on the
project, and chipped in a yearly hundred-scudo stipend.

In further correspondence with Vienna, Kircher cited a tremor in
his right hand as an additional challenge to completing the work. He
was approaching fifty. It's not known if this tremor was the result of
a stroke, of an old injury he suffered climbing cliffs and volcanoes, or
of having scratched his obsessive script over thousands of pages of

parchment. It's also not clear how serious the problem really was, since he finished off thousands and thousands more pages in his own hand over the next twenty-eight or twenty-nine years, before the tremor became totally debilitating. But it was reason enough to request a dedicated assistant to help with the manuscript. Kircher may have hoped all along to send for his friend Kaspar Schott. In any event, Ferdinand agreed to pay, and Schott was brought to Rome from Palermo.

It's possible to imagine that Kircher rarely thought of himself as a refugee from the war in Germany anymore. But it's more likely that memories from those early years never left him. And they probably contributed to the conceptual all-inclusiveness, the desire to bring everything together in harmony, that so often showed in his work. The war caused "the devastation of my entire fatherland," he wrote in his autobiography. And he was basically right, since it led to the death of about a third of the entire population. (That's an average. In many places it was much more.) In 1648, after years of negotiations, preliminary truces, and various settlements between the Swedes and the Saxons, the French and the Bavarians, the Spanish and the Dutch, as well as bargaining by the Protestant German states and a final Swedish siege of Prague, Kircher's patron, Ferdinand III, agreed to the Peace of Westphalia. And that's when the series of hellish episodes that began three decades before could finally be named. Later that very year, a German pamphlet appeared with the title *A Short Chronicle of the Thirty Years German War.*

But when it ended, the French were still in an ongoing war with Spain and on the verge of a civil revolt called the Fronde. The Dutch were still at war with the Portuguese. The Venetians were still at war with the Turks over Crete. The English were still at war with themselves, and on the way to beheading their king. The Swedes got most of what they wanted at the negotiating table, so one person who might be called a winner was twenty-one-year-old Queen Christina of Sweden. She was crowned at age six after the battlefield death of her father, Gustavus Adolphus, whose armies forced Kircher, Schott, and other Jesuits to run from Würzburg.

There was some confusion about Christina's gender at the time of her birth. (Questions lingered after her death, so much so that her remains were exhumed in 1966, but experts say the female skeleton they found doesn't rule out the possibility of an intersex condition.) Rumors about her sexuality were fed by her preference for men's coats and shoes, and by her appreciation for beautiful girls. It was reported that she "could shoot a hare with a single shot better than any man in Sweden." She thought of herself as an intellectual; she corresponded with Gassendi, read Tacitus, and enjoyed judging the quality of canvases, sculptures, and antiquities that came in crates as war booty from German lands.

The year after the Peace of Westphalia was signed, Christina decided that she wanted René Descartes to come to Stockholm and serve as her personal tutor. He was wary about doing it, and by the time he arrived from the Netherlands in late 1649, she'd become less interested in new philosophy than in ancient occult wisdom. Descartes had his hair curled for their first meeting, but she was disappointed in his looks, and didn't show much interest in him other than insisting he write a libretto for a court ballet. He barely got out of performing in it. She finally scheduled time for him: three meetings a week at five o'clock in the morning in her freezing-cold library.

(Even his thoughts froze in Sweden, he said.) Descartes caught the flu, which he treated by drinking hot brewed tobacco, and then developed pneumonia. Christina's physicians bled him, but naturally his condition grew worse, and after ten days of illness, Descartes died in Stockholm at age fifty-three.

But Descartes, who held on to his Catholicism while dismantling almost everything else, did have some influence on Christina, who was straining against Sweden's state-sponsored Lutheranism. Sometime after he died, Christina accepted visits from two Jesuits with long beards, disguised as Italian gentlemen. In secret meetings with these two, she laughed at literal belief in the Scriptures and at the possible apocalyptic import of a much-discussed comet. While she was less pious than she might have been for someone who was thinking about converting to their religion, her interest in Catholicism was at least partly genuine. In her view, the Jesuits encouraged, or permitted, the kind of intellectual curiosity she herself felt. She asked about astronomy, occult sciences, atomism, and, to one degree or another, about Athanasius Kircher.

Kircher lost no time contacting her, sending her a number of unctuous letters and his book on music. The queen answered: "I hope that we shall henceforth have the opportunity of greater freedom and sincerity to correspond with each other and to communicate with each other in greater safety."

12

Egyptian Oedipus

Now that he'd secured the emperor's funds and Schott's help, the onus was on Kircher to complete the monumental project he'd promised for so long. As Schott described it, Kircher's cubiculum during this period was piled high with manuscripts and books so old that many were "half-consumed by rot" or "besieged by dust and cockroaches" and "etched over in nearly illegible characters." Also lying around as they worked: "a vast quantity of hieroglyphs, idols, images, gems, amulets, periapts, talismans, stones, and similar things." Because *Egyptian Oedipus* was too ambitious and too long to be printed in one volume at any one time, the handwritten sheets had to be sent to the typesetter in several stages between 1652 and 1654. The book, all two thousand pages of it, was finally published in four volumes in 1655.

In his introductory treatise, Kircher likened his breakthrough with hieroglyphics to the discovery of America, the development of the printing press, and the sighting of "new heavenly bodies." Interest in Kircher's solution to the puzzle was already such that one Dutch bookseller alone bought five hundred copies (half the print run) of *Pamphilian Obelisk*, its 1650 precursor. But needless to say, it was not the kind of book that could be digested in a few sittings. As with most publications of the era, the assessments of readers emerged over a period of years and even decades, rather than weeks or months. Initially, especially among those who already thought of Kircher as "an oracle," it inspired awe. Much more recently it has been called "one of the most learned monstrosities of all times."

Egyptian Oedipus was full of esoteric-looking languages in exotic typefaces; long tables of obscure letters, markings, and symbols; occult diagrams; and engravings of mummies and pagan gods. The physical scale and the elaborate beauty of this book made it suitable for courtly display, consistent with the realization of a decades-long ambition on Kircher's part, and appropriate to what it was supposed to contain. *Egyptian Oedipus* wasn't just supposed to supply a method for retrieving the secret wisdom that was encoded in hieroglyphic inscriptions—it was supposed to provide the secret wisdom itself.

In Kircher's version of events, this wisdom had been handed down from God to Adam and passed along all the way to Noah. Over those centuries, impure versions of the sacred doctrines emerged in the form of dark magic, idolatry, and superstition, thanks in great part to Adam's bad son, Cain. After the flood, as Noah's descendants repopulated the earth—all the cultures of the world had to come from this same single source, after all—Noah's own bad son, Ham, made the great mistake of confusing the strains before going on to father the civilization of the Egyptians, who absorbed these idolatrous admixtures. It was Hermes Trismegistus, the thrice-great Egyptian priest, philosopher, and king, who was supposed to have decontaminated the sacred doctrines, at least to the degree that his

*Egyptian pyramids, as envisioned by Kircher
and his engraver*

pagan mind-set allowed, devising the obelisks and the system of hieroglyphics as a way to preserve them.

Superstitious practices crept back into Egyptian culture over time, Kircher argued, recombining with some true wisdom and forming the basis for various religions and societies. He thought he recognized ancestral Egyptian influence in the cultures of the Chinese, the Japanese, the Indians, and even the Aztecs, whose inscriptions and monuments he'd learned about through Jesuits in Mexico. Even Kircher had to admit it was all fairly complex. Putting together a scholarly chronology of human civilization that was consistent with the short time frame of the Bible, for example, presented challenges, though he betrayed little doubt of overcoming them. And there were many unknowns. On the question of influence between the Egyptians and their longtime captives the Jews (the descendants of Noah's son Shem), especially with respect to the mystical practices of Kabbalah, he had this to say: "I am fully persuaded that either the Egyptians were Hebraicizing or the Hebrews were Egypticizing."

And yet Kircher and many others believed that a strain of true (or truer) wisdom had survived, in the form of the written texts attributed to Hermes and in teachings believed to connect Hermes, Orpheus, Pythagoras, and Plato, among others. Kircher argued that vestiges of the hieroglyphic doctrine and symbolism must still lie hidden "scattered among the chronicles of ancient authors." So in order to decode the system, it wasn't just the "mysteries of the Egyptians" that he needed to apprehend, but also the "secrets of the Greeks, the amulets of the gnostics, the arcana of the Cabbalists, the phylacteries of the Arabs, the antidotes of the Saracen," and also the "characters, signs, frivolities, superstitions, and deceptions of all the imposters."

BEYOND THE CITATIONS of praise to Ferdinand in twenty-seven different languages, most of the beginning of *Egyptian Oedipus* was devoted to assuring readers how singularly suited Kircher was to be its author. Pages and pages of prose testifying to his erudition were preceded by lines and lines of poetry testifying to it. (Just who was this new Oedipus again? "Kircher's he.") In a foreword to readers, Schott described Kircher's superhuman diligence as a researcher and his uncanny ease with language: "He has been exceedingly educated in Arabic, Chaldaean, Syrian, Armenian, Samarian, Coptic . . . and for many years now he has spoken not only with Greeks in Greek, with learned Hebrew rabbis in Hebrew, but with Arabs in Arabic and with foreigners from other provinces of Asia and Africa, of which the number here in Rome at any time is enormous, in each one's mother tongue."

In a sense the entire first fifteen hundred pages of the book served as a kind of preface to Kircher's long-awaited translations. His stated strategy for deciphering them, a search for clues and shared symbols across antique texts and cultures, gave him an excuse to delve into the full spectrum of ancient Egyptian history, culture, and belief, and into every mystical tradition of the ancients that might contain some fragment of the original faith. There are extremely lengthy considerations of, for example, the metaphysical significance of numbers, mathematical harmonies, astrology, talismans, the musical magic inherent in the hymns of Orpheus, and the universal schemes of the Chaldean Oracles.

In many of these ideas he found commonality with the Hermetic texts he believed to be of Egyptian origin. In others he claimed to find fault. It wasn't that Kircher didn't believe in the mystical meaning of numbers or characters, or in cosmic sympathies and influences—in the radiation of divine forces from above and in invisible chains or "secret knots" linking intellectual, celestial, and earthly realms. These were the kinds of forces described in the texts supposedly written by Hermes Trismegistus, and that Kircher believed to be at work in magnetic attraction and repulsion. Generally, it was their corruption and use for divinatory, idolatrous, and superstitious practices that he objected to. Astral influence might be real enough, for example, but the practice of astrology for the purposes of telling the future amounted to black magic.

But while he claimed to disapprove of these pursuits, he frequently gave his readers what they needed to engage in them. "Long sections of the *Oedipus* described illicit magical practices with the detail of an instruction manual," writes the historian Stolzenberg. This thoroughness, for lack of a better word, made *Egyptian Oedipus* one of the biggest compendia of the occult ever produced, one that, in serving as a source for would-be mystics, had more influence on readers, and the culture at large, than Kircher's dubious translations of hieroglyphics ever would.

It also got him into trouble with the censors. Conservative members of the Jesuit College of Revisors, who were charged with upholding long-standing Church doctrines, were concerned by Kircher's manuscript. They ordered him to "explain doubtful things clearly, condemn blameworthy things," and "not assert magical or superstitious matters in detail." In the case of his extensive section on Kabbalah, where he cited "too respectfully the Talmud and other Jews," he was instructed to cut his discussion of the Sefirot way down, and the chapter on "practical Kabbalah" (including the use of numerical and alphabetical permutations for incantatory purposes) completely.

The censors could be forgiven for failing to understand exactly what Kircher was doing, and what he really believed. People have been wondering about that for centuries. But, among other desires, he wanted to reveal the way in which all people and all religious traditions were connected by (what he thought was) their common origin. Perhaps if you could trace the paths by which all the pagan and heretical peoples and cultures had gone astray, you could show the way back home, to the one true church. The Renaissance Neoplatonist Pico della Mirandola had decided long before that Kabbalah was a kind of Jewish version of the Hermetic doctrine and an ancient antecedent of Christian truth, as well as a perfectly decent vehicle for natural magic. The censors apparently didn't realize that Kircher's text on Kabbalah continued the revisionist Christianizing trend. He argued, for example, that the sefirotic tree of life, contemplated by Jewish scholars for ages, actually contained a notion of God as trinity. He'd adjusted the diagram of the tree a bit to reflect this.

In the end, with the approval of the superior general, the censors' requests were largely ignored. The reasons no doubt have to do with the power and prestige of Kircher's patrons, if not with a greater appreciation for his ideas. To comply with some of the concerns, Kircher did insert more of his perfunctory disclaimers into the text. The way in which these disavowals were mixed into otherwise sympathetic discussions was, in modern terms, "Jesuitical," and Kircher's contortions, in this book and others, surely contributed to that meaning of the word.

————

KIRCHER RELIED ON a daunting range of sources, and cited hundreds and hundreds of authors in a variety of languages. Prominent among them, Hermes Trismegistus, Neoplatonists such as Pico and Ficino, and one especially authoritative source, the Babylonian Barachias Nephi—the alleged, obscure author of the Arabic text that Peiresc and Barberini had been so keen for Kircher to translate all those years before. He had never quite gotten around to completing that project. This Nephi was nevertheless particularly loquacious on Egyptian notions of the divine mind and the power of the sun as a provider of "life, motion, and fecundity," as well as the idea that the Egyptians, too, worshipped a holy trinity. His statements often support Kircher's arguments so perfectly that it's believed Kircher wrote many of them himself.

Does it need to be said? Athanasius Kircher was an incredibly erudite man, but *Egyptian Oedipus* was not quite the work of erudition and scholarship that he made it out to be. Although standards of scholarship were much different than they are today, he frequently relied on the works of others when he claimed to have penetrated some obscure and exotic text himself, and he often entered into "full plagiarist mode," as one historian called it, for pages and pages at a time.

Readers who made it to the final tome of *Egyptian Oedipus* were at last provided with his interpretations of the (non-ancient) Bembine Tablet, a number of obelisks, and other inscriptions. Despite the years of work Kircher put into investigating the connection between Coptic and the language of the ancient Egyptians, he didn't spend a lot of time on a phonetic approach—what would turn out to be the correct approach—to interpreting the hieroglyphs. He had always been inclined to see them, as he had previously written, "not so much as writing but rather as symbolic representations of sublime theosophy expressed through signs that are universally intelligible." Universally intelligible, that is, to a select few with rare intellect and a divine calling.

Because he believed that Hermes Trismegistus had himself embedded the old doctrines in the hieroglyphs, Kircher's interpretations were bound to read like the other writings attributed to him—with perhaps an extra dash of pagan exoticism thrown in. Kircher's translation of a very small section of the Pamphilian obelisk, now at the center of Piazza Navona, is enough to get a sense of his work:

> The beneficent Being who presides over reproduction, who enjoys heavenly dominion . . . commits the atmosphere by means of Mophtha, the beneficent principle of atmospheric humidity unto Ammon most powerful over the lower parts of the world, who, by means of an image and appropriate ceremonies, is drawn to the exercising of his power.

That section is now known to contain merely the name and title of the Roman emperor Domitian—and to show that the obelisk was a first-century Roman commission and not, as Kircher claimed, a structure dating back to the fourteenth century before Christ. It was this sort of "flight of the imagination and learning run mad" that convinced one nineteenth-century Englishman that he could "safely consign" the "folio volumes of Father Kircher to the old book-stalls in Holborn."

When Kircher began his hieroglyphic studies so many years before, he had no idea that they were based on an incorrect assumption. This was now what he refused to believe. By the time Kircher began writing *Egyptian Oedipus*, he was well aware of scholarly research published in 1614 by the Swiss-born Calvinist Isaac Casaubon. Casaubon had dated the Hermetic texts to the second or third century, not to the time of Moses. There was no such person as Hermes Trismegistus, or, if there was, he was not the author of the writings and the hieroglyphic texts ascribed to him.

It's not really a surprise that Kircher wasn't willing to dismantle his sense of the sanctity of the Hermetic texts. That would have been like dismantling his sense of self, or at any rate, his life's work. On the contrary, he argued in *Egyptian Oedipus* that the kind of challenge to ancient authority that this Protestant Casaubon had made was a very dangerous thing: "Since they have been accepted by everyone, from so many centuries ago until these times," certain historical and sacred texts "are worthy of necessary trust."

"Without this the acts of all human affairs . . . would dissolve," he continued, and "nothing certain could be written or said, and all would be murky and obscured by doubt."

From a later perspective, Kircher employed some unscrupulous scholarly tactics in *Egyptian Oedipus*, but he was more concerned with what he saw as a noble intent. As a historian of the baroque period has put it, the goal was to show the "fundamental unity of human culture and its origins" as well as the fundamental truth of Christianity. For Kircher, larger truths took precedence over smaller ones. In the hands of someone like Casaubon, Kircher seemed to suggest, even the sanctity of biblical scripture might be called into question.

13

The Admiration of the Ignorant

Meanwhile—that is, while he was busy capping off two decades of work by producing what would later turn out to be wildly inaccurate translations of hieroglyphic inscriptions—Kircher also put together his famous museum.

During the last half of the sixteenth century, cabinets of curiosity, or wonder cabinets, had become popular all over Europe. Galileo had mocked the "curious little men" who displayed, say, "a petrified crab, a desiccated chameleon, a fly or spider in gelatin or amber, those small clay figurines, supposedly found in ancient Egyptian burial chambers." But his was the minority opinion. In 1651, when a wealthy Roman gentleman donated his trove of art and antiquities to the Collegio Romano, the Jesuits saw an opportunity to build a collection of prominence, and they put Kircher in charge of it.

The bequest included Roman, Greek, Egyptian, and Etruscan items—coins, statues, tablets, manuscripts, and more everyday artifacts from ancient or at least former Rome (household utensils, weights and measures). This gave Kircher a foundation on which to add all the curiosities he'd kept on view in his cubiculum. "It happened," he later recalled, "that I was compelled to transfer my private Museum to a more appropriate and accessible location in the Roman College, which they call the Gallery." This gallery, a long corridor adjacent to the formal library on the third floor, where paintings and portraits had already been hung, was well lit with three window bays.

Fairly soon Kircher and his assistants had filled it with swords, clothes, ornaments, preserved animal specimens, and other things brought or sent back by Jesuit missionaries. The collection included the "tail and bones" of a mermaid, which Kircher told visitors he obtained on Malta, and a brick from the Tower of Babel donated by Pietro della Valle. One Jesuit priest even contributed his own ten-ounce gallstone.

According to a catalog published years later, there were "armillary spheres, and celestial and terrestrial globes, equipped with their meridians and pivots." There were Archimedean screws, a number of mechanical clocks, and devices "bearing a resemblance to perpetual motion." There was "an organ, driven by an automatic drum, playing a concert of every kind of birdsong, and sustaining in mid-air a spherical globe, continually buffeted by the force of the wind."

Together, Kircher and Schott built a number of magnetic and hydraulic machines for the space. Some taught moral lessons, and some underscored traditional Aristotelian laws (for example, that nature abhorred a vacuum, despite recent experimental evidence to the contrary on the part of Evangelista Torricelli). Others were meant to reveal the wonders of natural magic, and a number of them "vomited" fluid, apparently to provide amusement that now seems inscrutably baroque. These included "a two-headed Imperial Eagle vomiting water copiously from the depths of its gullets"; "a hydraulic machine,

The Delphic Oracle

which supports a crystal goblet, from one side of which a thirsty bird drinks up water that a snake re-vomits from the other side while opening its mouth"; "a water-vomiting hydraulic machine, at the top of which stands a figure vomiting up various liquids for guests to drink."

Kircher's playful nature was never more in evidence than in his museum. He entertained visitors with the catoptric theater that John Evelyn had seen, the one in which cats were fascinated to observe "an innumerable multitude" of other cats, and with his projection device, which showed "ghosts in the air." He also delighted them with what he called the Delphic Oracle. To make it, Kircher removed the acoustical tube from the wall of his cubiculum and installed it in a similar recess between the college courtyard and the museum; it was then connected by progressively smaller hidden tubes to a hollow statue.

Let a "statue be situated in a sure and calculated spot in such a way that the end of the tube meets precisely with the opening of the mouth and you will have a statue perfect and consummate in articulately producing whatever you will," he instructed. "Because the orifice of the shell meets with a public place, all the words of men coming from outside into the spiral tube produce themselves drawn within the mouth of the statue." As a result, the tube could be "employed in playful oracles and fictitious consultations with such artifice that not one of its witnesses was able to discern anything concerning its secret construction." The Delphic Oracle is "shown to visitors not without some suspicion of a latent demon by those who do not understand its mechanism, for the statue opens and closes its mouth in the manner of one speaking, it even moves its eyes."

THE MUSEUM WASN'T just a place for the "investigation of the learned," "the admiration of the ignorant and uncultured," and "the relaxation of Princes and Magnates," as Kircher liked to say. It was a venue in which Kircher could put himself on display. And in general, given his increasing status and celebrity, the Jesuit authorities were happy for him to do so. Schott's recollections suggest that Kircher was in his element when giving tours of the museum. Sometimes he was in such a buoyant mood that he would pull a sword down from the wall to demonstrate his thrust and parry.

Kircher began to exhibit his own books in the museum and, at some point—ostensibly to show how integral he was to the functioning of the Republic of Letters and the degree to which he was in contact with the best and most holy minds—he also began to exhibit his personal correspondence. Before he died in 1680, Kircher exchanged letters with more than seven hundred fifty people, many of them quite prominent. The letter writing was never done: there were dispatches from Jesuits to acknowledge and forward to other learned men; sycophantic letters to send to patrons; sycophantic letters requiring response; and never-ending requests for translations. "It can hardly be said," Schott wrote, "how many inscriptions, sacred, profane, superstitious, magical and even diabolical . . . have been brought to him from all the parts of the world, in order to be interpreted."

It wasn't always easy being a "master of one hundred arts," as Kircher had begun to call himself. But indications are that Kircher felt a certain satisfaction during this period of his life (his early fifties). At least some of it had to do with the presence of his friend Schott, who worked tirelessly on Kircher's behalf.

MANY OF THE LETTERS sent to Kircher around this time concern astronomy. A Jesuit named Hermann Crumbach, for example, sent observations he'd received from Malabar, in southern India. A linguist, Amatus de Chezaud, reported on the "most minute asteroids" he'd observed from Aleppo. Former student Nicolò Mascardi wrote to him about the comets he'd seen while sailing through the Strait of Magellan. Although not a great astronomer himself (he lacked the attention span, for one thing, and the mathematical rigor), Kircher forwarded a steady stream of data from his correspondents to astronomers such as Giovanni Battista Riccioli, a Jesuit in Bologna, and Johannes Hevelius, the young astronomer from Danzig he'd hosted in Avignon many years earlier. Hevelius had used the proceeds from his family's brewing business to set up an observatory on the roof of his home. In an attempt to clear away the distortions and halos produced by the refracting telescopes of the time, Hevelius kept pushing their focal lengths, using a sixty-foot telescope to make unprecedented observations of the surface of the moon. They were published in one volume in 1647, in the form of one hundred fifty steel engravings. (Years later, he built a telescope with a focal range more than twice as long.)

The astronomical evidence kept mounting in favor of a sun-centered system. Kircher had come out against the magnetic arguments for the Copernican configuration in *The Magnet*, but that was more than twenty years before. Since then, because the Church subscribed to it, he'd perfunctorily endorsed the so-called geoheliocentric arrangement of Tycho Brahe. Supporters of this cosmology essentially said: *It appears to be true that the planets revolve around the sun—but the sun and the stars still revolve around the Earth.*

Schott had been encouraging Kircher to write on the subject. But an insufficiently critical discussion of the new astronomy could get you in trouble with the Inquisition, if not burned at the stake, and Kircher had always made a point of steering clear of it. He wasn't the type to take on the topic without a change in political circumstances, the kind that might have transpired, for example, if one of his friends were elected pope. In April of 1655, after the death of Innocent X, one of them was.

Kircher knew Fabio Chigi from their time together on Malta, when he served as apostolic delegate. Since then, they'd kept up a correspondence based on shared interest in such things as sundials and recondite learning. Chigi, who took the name Alexander VII, had risen to become the Vatican's highest-ranking diplomat, and was somewhat meekly involved in the negotiations that led to the Peace of Westphalia. As Kircher described it, Chigi was "finally exalted to the supreme tip of the apostolic peak by his altogether deserving and in no way ill-matched heap of merit." In reality, he was a compromise candidate whose election ended an eighty-day enclave.

Alexander VII was a strange little man who did "not enjoy what one would call perfect health," wrote a Venetian ambassador. "He is left with so few teeth that if he did not compensate the loss with false ones he would mumble." Slight and somewhat elfin in appearance, with an upturned mustache and a chin beard, he instituted his own particular brand of Vatican reform, renouncing papal nepotism and, at first, keeping his relatives away from Rome. Alexander "had so taken upon him the profession of an evangelical life," recalled a canon of Canterbury assigned to Rome, "that he was wont to season his meat with ashes, to sleep upon a hard couch, to hate riches, glory, and pomp, taking a great pleasure to give audience to ambassadors in a chamber full of dead men's sculls, and in the sight of his coffin, which stood there to put him in mind of his death."

The new pope wasn't very interested in affairs of state, but he cared about aesthetics. Although austere, he "liked his company to be gay in reason, and he enjoyed his intercourse." In frequent pain from kidney stones, Alexander would "speak quietly upon literature, ecclesiastical history . . . and upon the sacred sciences" such as Kabbalah and astrology and presumably also the new sciences. He spent afternoons in his apartments in such "literary meetings" with Kircher and others, and he "wished to have Bernini with him every day" to discuss architecture and urban-planning projects. Among other things, their conversations produced Bernini's design for the monumental colonnades that now form St. Peter's Square.

Alexander kept his favorites close, and it's possible to imagine that, for one reason or another, he was irked by Kircher's relationship with the less sophisticated, German-speaking Schott. There's no point in speculating further, but within months of Alexander's election, the Society of Jesus sent Schott all the way back to Germany—first Mainz, and then Würzburg. It was an "abrupt and unanticipated departure," says one historian, and was known to have caused both Kircher and Schott deep disappointment.

Whatever was said in the pope's apartments about Copernican theory, it was also within months of Alexander's election that Kircher moved forward with a book on the structure of the universe. But Kircher was under no illusions that he could simply ignore the sensitivities of the subject. Moreover, a new Jesuit ordinance included a prohibition on the discussion of magnetism ("action at a distance") with respect to cosmology. So he wrote it as a work of the imagination—the story of a cosmic dream in which an angel named Cosmiel leads Kircher's fictional stand-in, a priest named Theodidactus ("taught by God"), on an edifying flight through the heavens.

This wasn't the world's first space travel story. More than seventeen hundred years before, for example, Cicero had described a voyage among the stars in *Somnium Scipionis* (*The Dream of Scipio*). And Johannes Kepler's *Somnium* (*The Dream*), published a year after his death, in 1634, had taken readers on a dream visit to the moon. In Kepler's imagination, it was populated by "Privolvans," nocturnal creatures with legs like a camel's, skin like a serpent's, and "no established domicile." But Kircher's *Ecstatic Journey*, published in 1656, represented another step toward modern science fiction.

The book officially, ostensibly, argued for the system of Tycho Brahe, but it debunked many old Aristotelian notions. "You are mistaken, and greatly so, if you persuade yourself that Aristotle has entirely told the truth about the nature of the supreme bodies," Cosmiel explained. "It is impossible that the philosophers, who insist upon their ideas alone and repudiate experiments, can conclude anything about the natural constitution of the solid world, for we observe that human thoughts, unless they are based on experiments, often wander as far from the truth as the earth is distant from the moon."

The angel and the priest fly to the moon, which turns out to be cratered and mountainous and half covered by dark seas, and then to the sun, which is anything but a perfect sphere. (Cosmiel protects Theodidactus from the impossible heat by pouring "a vessel full of celestial dew" over his head. Later, in order to explore the sun's "immense Ocean, boiling with fervor," Cosmiel paddles him around in a rowboat made of asbestos.) Mars is "harsh with bulges, and blunt, and formidable with violent discharge of vapor." Jupiter is encircled by its recently discovered moons, and Saturn has "horrendous form."

As they head toward the stars, Theodidactus expects a collision with the crystalline celestial sphere . . . that never comes. Contrary to Aristotle, the stars turn out not to be fixed, and the universe appears to be boundless. But Cosmiel assures Theodidactus that it only *seems* infinite. (Giordano Bruno's belief in an infinite universe helped lead to his execution in the Campo de' Fiori in 1600.) It was all within God's realm. According to Scripture, God alone could count the stars.

THERE ISN'T MUCH DOUBT that Kircher privately believed in the Copernican model, but his opinion wasn't based solely on the astronomical evidence. A sun-centered universe also made much more mystical sense. In lieu of emphasizing magnetism in *Ecstatic Journey*, Kircher chose to emphasize something he called *panspermia*, or universal sperm. But the effect was the same: "The whole mass of this solar globe is imbued . . . with a certain universal seminal power," Cosmiel explains about the sun. It "touches things below by radiant diffusion."

When the Dutch astronomer and polymath Christiaan Huygens read *Ecstatic Journey*, he found it "for the most part inane and devoid of reason." But to other readers, in Europe and around the world, the book was "the offspring of consummate scholarship," as a Jesuit missionary in New Spain put it. In his view, Kircher was—forget Kepler, Galileo, Descartes, Huygens—"easily the Phoenix amongst the learned men of this century."

The Jesuit censors, for their part, were again displeased. "To be sure, Kircher on occasion reproves the condemned opinion of Copernicus about the motion of the Earth," said a report filed with Society authorities. "Nonetheless, throughout his entire book he carefully constructs all the evidence that Copernicus first brought in to establish and defend the motion of the Earth, and he weakens all the arguments by which that error is usually refuted."

"He may reject the motion of the Earth . . . and impugn it," the censors added, "but he does it so poorly."

———————

MORE ATTENTION MIGHT have been paid to the censors' concerns if the court of Rome hadn't been so distracted, caught up with talk about the impending arrival of Christina of Sweden. But Rome *was* distracted and caught up. The Protestant queen had abdicated her crown and converted to the faith of her defeated enemies. She was also rumored to be a hermaphrodite, or a lesbian, and to have killed off René Descartes.

Christina worked out the details of her conversion and her move to Rome in letters to the Vatican that were written in codes she herself had devised. For some time she had intended to abdicate, partly to be relieved of the pressure to marry and produce an heir. (She said she couldn't imagine being used by a man "the way a peasant uses his fields.") Given that she'd previously described religion as "a political invention for the restriction of common people," her reasons for converting were somewhat cynical. After her abdication ceremony in 1654, she rode out of Sweden on horseback, cutting off her hair and changing into trousers and boots before entering Denmark. Now, after a year and a half of ersatz royal residencies in Antwerp, Brussels, and Innsbruck, she was finally about to make her arrival in Rome.

The pope in particular obsessed over the details of her welcome. In his former role as chief papal diplomat he'd played some part in her momentous decision—a decision that, for all anyone knew, might lead to the end of heresy altogether. Chigi put Bernini in charge of decorating the Porta del Popolo, through which Christina would pass, as well as a matching fleet of vehicles for her use: a carriage, sedan chair, and litter (for reclining transport) all in sky-blue velvet and silver. In order to move her into the Palazzo Farnese, near

Campo de' Fiori, they kicked out another famous convert who had been living there for almost twenty years—the German prince, now a cardinal, with whom Kircher had traveled to Malta as father confessor.

Kircher shared a desire with the pope to win over Christina and gain access to her generosity. She had said in letters that she wanted to see her name prominently displayed in Kircher's books, and so Kircher had eagerly dedicated *Ecstatic Journey* to her. She was a collector of antiquities and a patron of the arts. Everyone assumed she had a fortune at her disposal and was getting ready to bestow it on Catholic Rome.

The celebration of her formal arrival, which began two days before Christmas in 1655, featured impossible throngs, an "interminable" procession through the city, and the entire College of Cardinals, in magenta robes, on mules, as well as cannon blasts from the Castel Sant'Angelo, chants in St. Peter's, dishes of gilded aspic, and bonfires in the Piazza Farnese. Some time after, she made a formal visit to the Collegio Romano, and later a less formal one. On her behalf, as one account has it, "Father Athanasius Kircherus the great Mathematician had prepared many curious and remarkable things."

It's hard to imagine what kind of rapport a bowing and scraping Kircher could have had with this queen in her man's cloak, black wig, and heavy face cream. She spoke a number of languages and, despite her interest in the occult, was not naive. One French scholar who visited her court in Stockholm wrote to Gassendi: "She has seen everything, she has read everything, she knows everything." Kircher showed her "the fountains and clocks, which, by vertue of the loadstone turn about with secret force." She also "stayed some time to consider the herb called Phoenix," which grew "perpetually out of its own ashes." It's not entirely clear what this herb called phoenix was, or what Kircher claimed—whether this plant was really supposed to have grown again from its "ashes" or, as he suggested in a later text, an image of the plant, or apparition, had been generated by exposing the plant's calcified salts to sunlight, moonlight, and then heat, thereby stimulating the plant's "seminal virtue" to re-form itself. It sounds most of all like one of Kircher's playful optical tricks.

Christina's visit to Kircher's museum increased its notoriety within Europe's elite circles, but it's not clear that she herself was all that impressed. Her approval didn't matter anyway, because in the days and weeks that followed, the hopes attached to Christina's arrival were largely dashed. She was known to talk all through Mass, she had the fig leaves removed from the nude statues in the Palazzo Farnese, and, it soon became clear, she had no money. (Still waiting for promised funds from Sweden, she'd been reduced to pawning jewels and silver plate.) On top of that, despite previous gossip about her lesbianism, it was said she'd begun an affair with the cardinal assigned to be her daily guide.

Celebrations for Christina nevertheless continued right on into Carnival season, in the weeks before the start of Lent. At night, mock battles between knights and Amazons in orange headdresses were waged in her honor. During the day, the queen watched the annual events on the Corso. Rome raced horses, donkeys, old men, boys, prostitutes, and Jews, at whom the crowds threw "everything from rotten fruit to dead cats."

Then, in the spring, the plague came.

14

Little Worms

It arrived in the Kingdom of Naples, apparently along with a transport of soldiers from Sardinia. At the worst point during the summer of 1656, thousands were dying in Naples each day. Desperate and frightened, citizens by the thousands flocked to the churches to pray for the plague to lift. According to one chronicle, people "of the highest quality," as well as the "disheveled," and presumably the infected, all joined these "confused processions," with the horrific result that "the streets and the stairs of the churches were filled with the dead." By the time it was over in August, as many as a hundred fifty thousand of the city's inhabitants had died. The epidemic began to ease around the day of the Assumption, August 15, lending credence to the notion that the Virgin Mary had finally interceded. The Jesuits claimed that their prayers to Saint Francis Xavier had made the difference. The Cistercians believed their prayers to Saint Bruno had.

Rome responded to the news from Naples by policing the seaports and the gates of the city. People and their animals were inspected. But by June the pestilence had "slithered" inside, as Kircher

put it, in this case via a fisherman off the boats at Nettuno. This
fisherman stayed at a rooming house in what were then the slums of
Trastevere, just across the Tiber from Rome proper, where he began
to feel ill. He died days later, "with evil signs." Possibly these were the
infected black buboes, or *bubones*—the term from which *bubonic* is
derived—swelling out as big as eggs or apples from under the arm-
pits and from the groin. The signs might have consisted of black or
red carbuncles over the entire body, infected and full of pus, or toward the end, a blackening of the skin from hemorrhage. If he'd been
coughing up bloody sputum, the disease had reached his lungs and
become pneumonic. Anyone who inhaled the spray in the air could
have been infected.

Was this the same plague responsible for the Great Mortality of
the fourteenth century, which killed off more than twenty-five million people, about a quarter of Europe's population? It's not perfectly clear what disease that was. The classic twentieth-century
explanation—that the bubonic plague bacterium (*Yersinia pestis*) was
carried by fleas, carried in turn by rats—has been challenged by a
theory that the disease may actually have been typhus or even an-
thrax, and not, or not only, bubonic plague. Nevertheless, between
1347 and 1670, *some* strain or form of pestilence attacked some part of
Europe every year except *two*.

As reported by Kircher, the sick certainly presented a confusing
and seemingly unlimited variety of symptoms. They developed not
only "boils and bumps, carbuncles and buboes of various forms," but
"harmful ear tumors or abscesses," after which followed a "loss of the
senses and unconsciousness, weakness and vomiting, hiccoughing or
coryza with fever." Sometimes "the patient begins to have fantasies
and talk wildly," though there might also be "loss of speech and de-

lirium, anxiety and pain about the heart, heat within the breast, distaste for all food, then follows often vomiting, fainting, straining of the heart, great thirst, heat and burning of the throat, sticking and lividity of the tongue, foul breath, frequent stools and severe nose bleed." Finally, usually within the week, "the poison rages within and conquers the entire body."

As soon as cases were confirmed in Trastevere, the Congregation of Health tried to contain the disease by sealing off the area. This work was done under the auspices of Kircher's former patron Cardinal Francesco Barberini; he had been allowed to return to Rome some years before, and now served on a kind of city health commission. Workers and soldiers came in the dark and built a wooden barrier around the district overnight, shutting in everyone who lived on its narrow medieval streets. Trade with Naples was discontinued, and great chains were dropped across the waterways to keep ships from sailing up the Tiber. Most of the city gates were shut, others were fortified with wooden stockades. But it was only a matter of days before more cases were reported in Trastevere, and then in the Jewish ghetto, and in other parts of Rome.

The plague was said to exist as "fetid miasma," or corruption of the air from putrid vapors. People thought that epidemics could be occasioned in the first place by celestial activity—a conjunction of malignant Mars with hot and humid Jupiter, for example. Other forms of corrupt air—due to decaying corpses, food, excrement, excessive humidity, stagnant water, emissions from volcanoes or other openings in the earth—were said to join with this miasma and make things worse. The poison could stick to clothes and bedsheets, sacks, cords, ribbons, and hair, and could penetrate the body through pores in the skin.

A plague doctor of Rome, about 1656

To counteract it, people scrubbed floors and walls with vinegar; burned rosemary, cypress, and juniper; and rubbed oils and essences on their skin. The wealthy left for the country if they could. Beggars were simply sent to prison or conscripted to help the sick and scrub the streets of rotten garbage and excrement. When members of middle-class households were found to be sick, their houses were often quarantined, with families boarded up inside. The vast majority of the ill, and sometimes those likely to become ill, were taken where their exhalations could do the least harm, to quarantined pesthouses, also called lazarettos, after the biblical story of Lazarus—though if you went in, the chances of dying, and staying dead, were high. "Here you are overwhelmed by intolerable smells," wrote a visitor to a lazaretto in Bologna some years before. "Here you cannot walk but among corpses. Here you feel naught but the constant horror of death."

In the middle of the Tiber, the entire island of San Bartolomeo, site of two churches, a convent, a hospital, and a number of houses, was transformed into Rome's main pesthouse. Over the two bridges from Rome and Trastevere, past double timber gates, hundreds, perhaps thousands suffered at any given time. Patients shared beds and received both religious and medical care. They were given special preparations of snake meat called theriaca, which because of its sympathetic (or magnetic) power was supposed to draw up the plague's poisons. The black buboes might be lanced and drained, or cut open (they were often infected and foul smelling) so that leeches could be applied. Sometimes the buboes were brought to a head with irritants such as rock salt or turpentine. The dead from the lazarettos were stripped of their clothing so it could be burned, then loaded on barges, taken down the river, away from the city, and dropped in pits deep enough, as Romans thought, to prevent the fetid, contagious miasma from rising out of the ground again.

Because of what they knew about the pesthouses, people with

signs of the plague tried to hide them, and people with dead relatives (the worst sign of all) tried to hide *them*. The sick purged and bled themselves, and lanced and drained and cauterized their own buboes and carbuncles, but were often dragged off to the island quarantine anyway.

In late June, schools, courts, markets, and businesses—almost everything except churches—were closed. A general lockdown of the city was ordered for forty days. Doctors walked the quiet streets dressed in seventeenth-century versions of hazmat suits: waxed robes, goggles, and beaklike masks stuffed with herbs and spices. They wore sponges soaked with vinegar around their necks and sometimes carried torches or buckets of burning tar to clean the air. Soldiers, policemen, and health officials patrolled the city. People who were found violating health regulations or quarantines were jailed, or in many cases put to death. One thirteen-year-old girl, who had run out into the street after a chicken, was hanged, apparently as a lesson to others. The Vatican made an effort to help the poor and the sick and to pay for lazarettos, but there was a food shortage in Trastevere and, since the city had been brought to a standstill, a loss of work and pay everywhere.

Although the "plague most atrocious" continued for more than a year, many of these tactics of "opportune suppression," as Kircher characterized them, did help prevent the spread of the disease. (Not the snake meat, presumably.) The effects of the epidemic were "considerably more mild than . . . at Naples"—only about fifteen thousand people died—and yet living through it was frightening. The "altogether horrid and unrelenting carnage" of Naples was on everyone's mind, Kircher remembered, and "each man, out of dread for the ever-looming image of death, was anxiously and solicitously seeking an antidote that would ensure recovery from so fierce an evil." As a melancholic, Kircher was supposed to be particularly susceptible.

"In this state of affairs," he remembered, "amidst the horrible silence of the sad city and in the deepest recess of solitude (for the entrance of the Roman College had been closed), I attempted with sluggish though necessary toil to develop the ideas that I had previously begun to conceive concerning the origin of the plague."

There was almost no commonly held, or seemingly contradictory, view of the plague to which Kircher gave short shrift. He believed that pestilence originated with God as a form of penance after the Great Flood (along with war, famine, and death), but that it raged and receded, and could be treated, by natural means. With respect to contagion, you could breathe it in. And if you ate food grown in soil that had become foul from the upward seepage of putrid poisons, you could ingest it. If you suffered from a corruption or obstruction of humors, apparently you could even generate the pestilential putrefaction yourself.

Most of Kircher's ideas weren't new. But during the course of his study, he became one of the very first people in history to use the microscope to study disease. Perhaps the very first. And he applied his findings to an argument that was either ancient or brand-new, or both, depending on how you looked at it or on the century from which you looked.

It is "generally known that worms grow from foul corpses," Kircher wrote in *Examination of the Plague*, published in early 1658. "But since the use of that remarkable discovery, the smicroscopus, or the so-called magnifying glass, it has been shown that *everything* putrid is filled with countless masses of small worms, which could not be seen with the naked eye and without lenses." Kircher would not have believed it himself if he hadn't seen it with his own eyes through what he believed was a very sensitive instrument. He'd performed an entire series of "experiments," which he invited readers (with access to a similar instrument) to try. For example:

> Take a piece of meat, and at night leave it exposed to the lunar moisture until the following day. Then examine it carefully with a smicroscopus and you will find that all the putridity drawn from the moon has been transformed into numberless little worms of different sizes, which in the absence of the smicroscopus you will be unable to detect. . . .

The same basic thing happens, he said, if you take a bowl of water sprinkled with dirt from the ground and expose it to the sun for a few days: "You will see . . . certain vesicles which are quickened into exceedingly minute worms" that eventually become "a vast number of winged gnats."

It also happens "if you cut a snake into little pieces, soak them in rain water, expose them for some days to the sun, bury them for a whole day and night in the earth, and then, when they are soft with putridity, examine them with a smicroscopus"—except in that case the decaying mess is swarming with little "snakes."

Kircher was hardly oblivious to the growing emphasis on experimentation, direct observation, and physical evidence during his own time. He frequently took pains to assure readers that only those ideas "which have been verified and proven by experiment" had been set down, and that those which were the "product only of opinion and remain unsupported" had been passed over. It's just that his understanding of what it meant to perform an experiment didn't include modern expectations related to scientific hypothesis and procedure. The word *experiment* comes from the Latin for "try" or "test," as in to try out an idea rather than just think about it, but in Kircher's case it didn't necessarily mean that the idea itself was to be challenged. On the basis of these and other "incontrovertible experiments," there was really no doubt in his mind about what he already thought he knew: that worms, insects, and similar creatures grew from the decay of other living things. And it occurred to Kircher that, at least as he and everyone else believed, the plague *also* arose from decay and putridity.

Kircher examined the blood of plague sufferers under his microscope on several occasions. "The putrid blood of those affected by fevers has fully convinced me," he wrote. "I have found it, an hour or so after letting, so crowded with worms as to well nigh dumbfound me." His assertion: "Plague is in general a living thing."

Kircher's arguments about spontaneous generation are hard to follow, largely because he did so much prevaricating and used so many ill-defined terms interchangeably. It was all utter speculation, anyway. Universal sperm was certainly involved: he believed that "seeds of a vegetative and sentient nature" are "scattered everywhere among the elemental bodies." Whether decomposing matter also generated life on its own or acted as a fertilizer or necessary ingredient is, as one historian commented, "not perfectly clear." In the case of the plague, Kircher suggested that when the tiny *semina*, or seeds, or "corpuscles," that emanate from all natural things are corrupted by putrescence, they become the minute carriers of the disease. "Corpuscles of this kind are commonly nonliving," he explained, "but through the agency of ambient heat already tainted with a similar pollution, they are transformed into a brood of countless invisible little worms."

Specifics aside, Kircher was making a larger, more fundamental argument against the new mechanistic, material philosophy of Descartes and others, based on which even animals and plants were nothing but elaborate machines. He refused to concede that the physical world was merely physical. To him it was animate on some very basic level.

According to Kircher, "these worms, propagators of the plague, are so small, so light, so subtle, that they elude any grasp of perception and can only be seen under the most powerful microscope." Therefore, they "are easily forced out through all the passages and pores" of plague victims' bodies and of the sick, and "are moved by even the faintest breath of air, just like so many dust particles in the sun." They are then "drawn through the breath and through the sweat pores of the body, from which later such fearful symptoms and effects result."

A MEDICAL HISTORIAN writing in 1932 described Kircher's *Examination of the Plague* as "a farrago of nonsensical speculation by a man possessed of neither scientific acumen nor medical instinct." But two years before, another historian determined from it that Kircher was "undoubtedly the first to state in explicit terms the doctrine of 'contagium vivum' as the cause of infectious disease"—in other words, that Kircher discovered microorganisms and was the first to propose the germ theory of contagion. If that's true, however, then his articulation of germ theory was predicated on notions (spontaneous generation, animism) that no modern scientist would be caught dead advancing. Besides, the concept of universal seeds went back to the Greek philosopher Anaxagoras, and the idea that disease is living turns out to be both ancient and mystical.

A lot of what Kircher wrote about plague came from Lucretius, the disciple of Epicurus from whom Gassendi got his modern-sounding ideas about atoms. In his epic poem *On the Nature of Things*, Lucretius wrote that "there are many seeds of things which support life, and on the other hand there must be many flying about which make for disease and death." They can come "down through the sky like clouds and mists, or often they gather together and rise from the earth itself, when through dampness it has become putrescent."

Among other sources, Kircher also borrowed heavily from, but doesn't cite, a sixteenth-century writer most famous for a verse trea-tise called *Syphilis, or the French Disease*. (Understandably, no one wanted to take the credit for syphilis, which reached epidemic pro-portions during this time. To Muslims, syphilis was the disease of the Christians, to the English it was the French pox, to the French it was the Neapolitan disease, and to the Italians it was of Spanish origin.) The author, a physician from Verona named Fracastoro, worked out a theory of contagion involving transmission of "imper-ceptible particles," infected and self-propagating, that he called *semi-naria*, or "seedbeds," sometimes translated as "germs."

What did Kircher actually *see* when he examined the blood of plague patients? He claimed his microscope made "everything ap-pear a thousand times larger than it really is," but he didn't mean that literally. It's not possible that he saw plague bacilli, which are one six-hundredth of a millimeter long. He may have been using a relatively simple magnifying lens rather than a compound microscope, which employs a system of lenses, in which case he would have been lucky to see much of anything. Even with a compound microscope, any organic specimen might have looked like a mass of tiny worms.

Yet most of Kircher's readers had never looked through a microscope at all. Only one or two treatises on the subject had ever been published, and so *Examination of the Plague* caused a kind of sensation within the Republic of Letters. A doctor in Dresden compared Kircher's brilliance to the shining of the sun. An anatomy professor in Jena informed Kircher that "the reputation of things Kircherian" had "spread through all of Europe." In thanks for his copy of the book, a missionary in New Spain sent Kircher chocolate and peppers "which we call Chile." Eventually, whether or not *Examination of the Plague* hit upon one of the core tenets of epidemiology, it did influence thinking about the way disease is actually spread.

Eager readers may have been under the incorrect assumption that Kircher had something to do with stemming the plague in Rome. And on the question of prevention and cure Kircher *did* provide his considered opinion, but in this case it can't remotely be mistaken for an early expression of modern medical thought: he believed that, short of leaving the area, an amulet made of the flesh of a toad or of dried toad powder, and worn over the heart, was probably the best antidote.

15

Philosophical Transactions

Three years after the plague subsided in Rome, the Tiber flooded, having the greatest effect on the Jewish ghetto. Alexander VII spent much of his time in the Palazzo Quirinale with his wooden scale model of the streets and buildings of the city, contemplating his improvement projects. Although unenthusiastic about matters of government, he was given respect for the success of Rome's efforts against the plague, and he had done his part in 1657 by sending the papal fleet to join Venetian ships at the Battle of the Dardanelles, part of the ongoing war with the Ottoman Empire over Crete. He'd also bowed to tradition with regard to nepotism; a little more than a year after his election his brother and nephews had gone on the payroll.

Despite Kircher's previous output, he had only begun to become, as a twenty-first-century historian has put it, "a book-making, knowledge-regurgitating machine." In 1658, he published *Ecstatic Journey II*, a precursor of a planned volume on the physical earth called *Underground World* that was going to take several more years to complete. In 1660, after the first major eruption of Mount Vesuvius in three decades, Kircher traveled down to Naples to investigate an apparent miracle. Crosses had mysteriously begun to appear in the folds of people's clothes, aprons, bed linens, and other fabrics. Kircher the skeptic determined that the causes were natural, not miraculous, a result of the ash in the air—though this didn't mean that God wasn't responsible for them. A book on the subject naturally followed.

Also on Kircher's agenda in the years after the plague: ingratiating himself with the new Holy Roman Emperor in Vienna. After Ferdinand III died in 1657, his seventeen-year-old son, Leopold, king of Bohemia and Hungary, succeeded him. The child of married first cousins, Leopold himself happened to be first cousins with Louis XIV of France. He was born with what became known as the Hapsburg chin—a greatly protruding jaw and almost monstrous lower lip—the result of so much inbreeding among European royalty. (His nickname: Hogmouth.) Like his father, and like other kings and queens whose subjects, peers, and family members spoke many different tongues, Leopold had a practical interest in crossing the divide of language. At the same time, because he frequently wanted to keep others from being able to read and understand his secret missives and official directives, he'd developed an interest in cryptography. For Leopold, Kircher produced both an attempt at a universal language that would allow any two parties to communicate and a system of artificial languages or codes by which only certain people could.

In fact, the possibility of a universal language was frequently discussed during the seventeenth century by the likes of Descartes, Leibniz, and many others. In the early 1650s, for example, the Englishman Francis Lodwick proposed one in a book called *The Groundwork or Foundation Laid (or So Intended) for the Framing of a New Perfect Language and a Universal Common Writing*. In 1657, another Englishman, Cave Beck, published his proposal in *Universal Character, By which all the Nations in the World may understand one another's Conceptions, Reading out of one Common Writing their own Mother Tongues*. Kircher's own attempt, set down in *New and Universal Polygraphy* and made available to a select readership beyond the emperor, was based on the use of the combinatory arts he was so fond of. In essence it was an elaborate system for translating more than a thousand core terms among five languages. It lent itself to the same kind of computing-box system as his method for composing music. As a result, Kircher made a number of wooden arks or "organs" for patrons containing many categorized, combinable slats. They served as tabletop or desktop aids to composition in his "universal language" and a variety of secret codes.

In Germany, Kaspar Schott dutifully continued to work on Kircher's behalf, among other things publishing a new edition of *Ecstatic Journey* meant to address the censors' concerns, and taking over an entire four-volume project on natural magic that Kircher had recognized he would never complete. Schott produced several other fat books of his own, on everything from "physical curiosities" to "technical curiosities" to "serious amusements." Their final project together involved another one of Kircher's seventeenth-century computers. He called it his mathematical organ, and it worked on the same principle as his other machines. The slats in this one were meant to enable calculations in every field in which a young prince might need to make them: arithmetic, geometry, fortifications, chronology, horography, astronomy, astrology, steganography, and music. Schott wrote an instruction book for it from his base in Würzburg, finishing just before he died of what must have been exhaustion in 1666. The manual, 850 pages long, was published posthumously.

BY 1661, Kircher's writings were so well known that Johann Jansson, a prominent Amsterdam publisher of atlases and other handsome volumes, approached him with a proposal for an agreement of international scope—an offer of twenty-two hundred scudi for the rights to publish his present and future books in several countries of continental Europe and in England. Not only did Kircher agree to the offer, he was known to mention it frequently in letters and conversation with others.

For these new Protestant publishers, Kircher decided to produce a volume on the history, topography, and landmarks of Latium (Lazio in Italian), the countryside around Rome. It was a subject that could be illustrated with many beautiful engravings, and one that was worthy of scholarly attention. According to the historian Ingrid Rowland, for example, there was a tradition of thinking that after the Flood, Noah "had taken to traveling the world by raft, vigorously repopulating the flood-drenched continents with the heroic aid of his wife." And one strain of Hebrew scholarship suggested that Noah had finally come to rest on the Italian peninsula. (Indeed, the fifteenth-century Dominican monk named Giovanni Nanni, aka Annius of Viterbo, spent his career forging—and then translating and commenting upon—ancient Chaldean, Etruscan, and Egyptian texts in order to prove that Noah had chosen to settle down in his own Italian hometown of Viterbo.)

So it was that later in 1661, Kircher traveled to the ancient town of Tibur "for the sake of refreshing my strength and at the same time to gather material on antiquities for my work on Latium." It was about thirty miles east of Rome, at the foot of the Sabine Hills. Now called Tivoli, Tibur had been a famous summer residence for ancient Romans, and was the site of an enormous villa built by the Emperor Hadrian. One day during his time there, Kircher set out in search of a particular ruin: "I had heard that somewhere in the nearby mountains there was hidden the famed rubble of the city of Empolitana, referred to rather often by Livy," he recalled.

"After procuring a travel mate I went out to survey the area, an utterly difficult journey." They hiked many hours, but the ruins of Empolitana proved elusive. So the two began a very long climb up the slope of the highest mountain in the small range called the Monti Prenestini, which would offer the best view of the hills and valleys on either side below. The mountain was originally called Mons Vulturum or Mons Vulturella, the Mountain of Falcons. Over many hundreds of years the name had become Mentorella. It was sometimes also known as Monte Guadagnolo, after the little village that clings to its southern slope. Above the village is a weathered ridge of scrub and rocks that runs along the length of the mountain. It's about three thousand feet above sea level, and fifteen hundred feet down to the green valley below. Little villages and towns are visible in the foothills of the range a couple of miles to the east. As Kircher and his companion made their way along this ridge, they saw signs of a building on the steep eastern face of the mountain.

The building turned out to be a church "nearly consumed with age." It stood on a sort of shelf against a tall, craggy outcrop. "When I drew nearer I realized that formerly it had been a magnificent structure," Kircher recalled. "Still, I marveled that it had been placed in this bristling wilderness."

Examining the remains of the church, he found a marble inscription: this was the site of the conversion of Saint Eustace. As the story went, sometime early in the second century, a Roman commander named Placidus was hunting on the mountain when a stag came out and stood at the top of the jutting rock. Appearing between its antlers was a vision of the crucifixion of Christ. This was enough to turn Placidus from paganism to Christianity and inspire him to take the name Eustace. After he refused to offer sacrifices to the pagan gods in Rome, however, he and his family were burned to death in a brass bull. Two hundred years later Constantine the Great supposedly came across the site where Eustace had seen Christ, and built a church there. (Today the entire story has been roundly discounted.)

Kircher went up to the altar of this dilapidated church, where there was a statue of the Blessed Virgin. It was "venerable in its antiquity" but "very neglected" and "covered by a filthy rag." And then "with a certain marvelous impulse from an inner spirit She was seeming to address me," he wrote. How "deserted by all in this horrible place" she was; how in the past she "was blooming abundantly with great devotion in this very spot."

Kircher was "thoroughly moved" to his "innermost viscera" by his experience. The view from high up on that mountain was like a more dramatic version of the hilltop view from Geisa. He returned to Rome determined to undertake a restoration of this very remote little church.

WHILE KIRCHER WAS up on the mountain, members of the newly formed Royal Society of London, an entity that has taken on legendary status for its role in the development of Western science, were trying to figure out how to contend with Kircher's work.

Some of the founding members of the Society had begun meeting less formally in the 1640s, in the midst of the English Civil War, and some at Oxford in the 1650s. (They were mostly Parliamentarians rather than Royalists, though the founding group was a mix of both.) After lectures at Gresham College in London, they got together in the lodgings of one professor or another, or in taverns such as the Mitre on Wood Street and the Bull Head in Cheapside, to talk about everything from the "Copernican Hypothosis," and the "weight of the Air" to "Valves in the Veins" and the "Descent of Heavy Bodies." One of them maintained an "operator" in his lodgings to grind glasses for telescopes and microscopes; they also had access to the chemicals and instruments of an apothecary.

As described in an early history of the Royal Society, most of the people in this fairly astonishing group were "gentlemen, free and unconfined," meaning that with "the freedom of their education, the plenty of their estates and the usual generosity of noble blood" they had the means and the wherewithal to pursue their interests. Collectively, they had studied Descartes, or met him in Paris, and engaged in correspondence with such people as Mersenne, Gassendi, Christiaan Huygens, and Hevelius. They admired William Gilbert's *De Magnete*, William Harvey's *De Motu Cordis* (*On the Motion of the Heart*), and Francis Bacon's vision for an entirely new approach to knowledge based on inductive reasoning and the experimental method. Many of them also had a connection to Kircher.

In a letter written in early 1651, Dr. William Petty—a physician, chemist, economic theorist, and professor of music who also directed a comprehensive land survey of Ireland, designed double-hulled boats, and later became a founding member of the Society— described the initial approach of the Oxford group: to try to establish what had been achieved to date within various areas of knowledge. "The Club-men," he wrote, in pre-standardized English, "have cantonized or are cantonizing their whole Academie to taske men to several imploiments and amongst others to make Medullas of all Authors in reference to experimental learning. Thus they intend to doe with Kircherus Workes and others." Where else would they go, frankly, for the most complete compendia of progress in so many fields?

The society was officially founded in 1660, after a lecture on astronomy by Christopher Wren, to promote "Physico-Mathematical Experimental Learning." Chairing the meeting: John Wilkins, a liberal clergyman who supported the heliocentric model of the universe and believed in the possibility of space travel. Author of a 1648 book called *Mathematicall Magick, or, The Wonders that May Be Performed by Mechanicall Geometry*, he also joined the quest for a universal language, publishing the results of his very elaborate but doomed effort in 1668. (His classification system later inspired a short essay by Jorge Luis Borges, which in turn inspired an entire book, *The Order of Things*, by Michel Foucault.) During the 1650s Wilkins hosted many meetings of what was sometimes called the Invisible College in his lodgings at Oxford. According to another early Royal Society fellow,

John Evelyn (who had documented the papal procession while visiting Rome in 1644), Wilkins put many "artificial, mathematical, and magical curiosities" on display in his room. These included a "monstrous magnet," a machine that made rainbows, and a speaking statue. The statue "gave a voice and uttered words by a long concealed pipe"—and sounds an awful lot like it was modeled on the one in Kircher's museum.

Sir Robert Moray, another founding member, had developed his interest in what would now be called scientific matters, as well as natural magic and the Hermetic tradition, by studying Kircher's *Magnes* during a seventeen-month imprisonment at the hands of the Duke of Bavaria in the early 1640s. Moray, a Scot and a Freemason who had worked as a spy for Cardinal Richelieu, was captured while serving in the Scots Guard with the French in the Thirty Years War. Moray had a lot of time on his hands and began a correspondence with Kircher that lasted decades. He later wrote to Kircher about the extreme tides of the Western Isles, and referred others to *Egyptian Oedipus*. (He also seems to have incorporated symbols and ideas from *Egyptian Oedipus* into the imagery and rituals of Freemasonry.) After working behind the scenes at a very high level to help effect the Restoration of the English monarchy and bring Charles II to power in 1660, Moray was given his own house to live in on the grounds of Whitehall Palace. He secured the charter from the king that gave the Royal Society its name.

Charles II was himself a dabbler; he outfitted a laboratory in Whitehall, from which the noise of mechanical instruments and the smell of chemical (or alchemical) experiments could often be detected. After a while, in foppish imitation of the king, his court and courtiers went from "baiting Puritans, place jobbing, flirting, and gambling," as one nineteenth-century writer put it, to "discussing the pneumatic engine, the ponderation of the air, blood transfusion, and the variations of the compass."

Links to Kircher were widespread among the Royal Society's members and their experiments. Robert Boyle, who in 1661 published *The Sceptical Chymist*, an attempt to sort Hermetic alchemical fictions from experimental chemical facts, is also known for his work on vacuums, atmospheric pressure, and the properties of air, conducted in the late 1650s. Only fifteen years or so before, the jury was still out on whether vacuums even existed. Kircher, obliged to deny the possibility of a vacuum (vacuums were abhorred by nature, per Aristotle), had been present at an inconclusive experiment involving a siphon, water, and a very long lead tube, conducted in Rome sometime in the early 1640s. Kircher disingenuously reported that it had failed. But that experiment helped inspire Evangelista Torricelli, who in 1644 not only created a vacuum but essentially invented the mercury barometer—and *that* experiment inspired a great deal of discussion and trial by Boyle and others. In 1657, two years after Kircher's friend Kaspar Schott returned to Germany, Schott published the first of his own books, an aggregation of information on mechanics, hydraulics, and pneumatics. He somewhat unenthusiastically included a report on the air pump recently invented by Otto von Guericke of Magdeburg, which Boyle read. Boyle and his assistant, Robert Hooke, made an improved version of it, which allowed them to carry out their unprecedented series of experiments, published in 1660. And so it wasn't Kircher but his disciple who helped put old notions about the impossibility of a vacuum to rest.

"Father Kircher is my particular friend, and I visit him in his gallery daily," Robert Southwell, who later became president of the society, wrote to Boyle while visiting Rome in 1661. "He is likewise one of the most naked and good men that I have seen, and is very easy to communicate whatever he knows. . . . On the other side he is reported very credulous, apt to put into print any strange, if plausible story that is brought unto him. He has often made me smile."

Conflicted feelings about Kircher were fairly common. In the late 1650s, before becoming the organization's secretary, Henry Oldenburg dutifully tried to get to the bottom of the vegetable phoenix that Kircher had put on display for Queen Christina in 1656, but failed. Later, as the editor of the *Philosophical Transactions of the Royal Society*, the world's first scientific journal, Oldenburg published long summaries of Kircher's new volumes when they became available. The problem was that attempts to reproduce his experiments frequently didn't succeed, and Kircher's claims and propositions frequently didn't hold up. No one at the time thought to give him credit for playing what can now be seen as a very valuable role: providing so many statements to test against, a means by which to determine what wasn't true.

If some of these Englishmen begrudged him, it wasn't for his fascination with magnetism. Despite the general success within England of new ideas, including the Cartesian notion that only material explanations of natural phenomena could hold weight, interest in this apparently immaterial power of attraction was still high. Christopher Wren believed it must be responsible for the motion of the planets. King Charles II was fascinated by magnetism and presented the Royal Society with a special terrella for use in its work. Boyle's own theological musings made use of the analogy of the lodestone to describe God's ways. And John Milton's *Paradise Lost*, begun in 1658 and published in 1667, contains language about the magnetic rays of the sun that might even be called Kircherian:

> . . . magnetic beam, that gently warms
> The universe and to each inward part,
> With gentle penetration, though unseen
> Shoots invisible virtue ev'n to the deep . . .

Wren, Boyle, Wilkins, and others certainly shared Kircher's fascination with the microscope. Their interest may have been spurred in the first place by discussions of the instrument in *The Great Art of Light and Shadow* and *Examination of the Plague*. But, starting in 1663, it was Robert Hooke, himself diminutive, and by then the Royal Society's curator of experiments, who really opened up the microscopic world. Hooke's book *Micrographia*, published in 1665, included observations and beautiful copperplate engravings of dozens of items, including, as Oldenburg's account in *Philosophical Transactions* described it:

> Edges of Rasors, Fine Lawn, Tabby, Watered Silks, Glass-canes, Glass-drops, Fiery Sparks, Fantastical Colours, Metalline Colours, the Figures of Sand, Gravel in Urine, Diamonds in Flints, Frozen Figures, the Kettering Stone, Charcoal, Wood and other Bodies petrified, the Pores of Cork, and of other substances, Vegetables growing on blighted Leaves, Blew mould and Mushromes, Sponges, and other Fibrous Bodies, Sea-weed, the Surfaces of some Leaves, the stinging points of a Nettle, Cowage, the Beard of a wild Oate, the seed of the

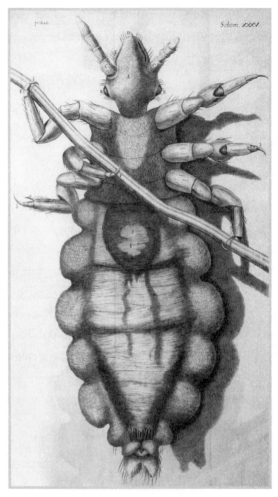

A louse, from Hooke's Micrographia

Corn-violet, as also of Tyme, Poppy and Purslane. . . . Hair, the scales of a Soal, the sting of a Bee, Feathers in general, and in particular those of Peacocks; the feet of Flies; and other Insects; the Wings and Head of a Fly; the Teeth of a Snail;

the Eggs of Silk-worms; the Blue Fly; a water Insect; the
Tufted Gnat; a White Moth; the Shepherds-spider; the Hunting Spider, the Ant; the wandring Mite; the Crab-like insect,
the Book-worm, the Flea, the Louse, Mites, Vine mites.

The book also included a few brief references to Kircher, which
may be seen as either an understated admission of his significant
influence or a true indication of his minor role.

A few months after *Micrographia* was published, the plague came
to London. From June 1665 to March 1666, the Royal Society's weekly
Wednesday meetings were canceled. As many as a hundred thousand people died. Then, in September 1666, the Great Fire of London burned for four days, destroying thirteen thousand houses and
eighty-four churches. About the churches: Christopher Wren, better
known today as an architect than as an astronomer, rebuilt fifty-one,
as well as St. Paul's Cathedral.

16

Underground World

J ust before Hooke published his microscopic observations of
everything from the "Edges of Rasors" to "Vine mites," Kircher
published *Mundus Subterraneus* (*Underground World*), a two-
volume tome of atlas-like dimensions, intended to lay out "before the
eyes of the curious reader all that is rare, exotic, and portentous contained in the fecund womb of Nature." This was the first of his books
to be printed by his eager new partners in Amsterdam, and while it
was traditionally deemed unworthy of natural philosophy to delve
below the surface of the earth, into its nether regions, Kircher believed it was all part of God's sometimes incomprehensible and yet
perfect creation, a dark realm whose relationship to the light he
wanted to explicate. There is an "idea of the earthly sphere that exists
in the divine mind," he proclaimed, and in this early work on geology
he tried to show that he had grasped it.

Kircher believed he was in a special position to reveal the hidden
world below. After all, he wrote, referring to himself, "the author was
present with great danger to his own life" during the horrible earth-

quakes of Calabria in 1638. It was during that time that he "learned the great secrets of Nature," and this firsthand experience provided the pseudo-empirical proof he needed to conclude—as Plato, Aristotle, Pliny, Vitruvius, Cicero, and many others assumed—that some type of central fire existed deep inside the earth.

In Kircher's view, volcanoes, however awful and awe-inspiring, "are nothing but the vent-holes, or breath-pipes of Nature." Earthquakes are merely the "proper effects of subterrestrial cumbustions" that are sure to go on constantly. The "prodigious volcanoes and fire-vomiting mountains visible in the external surface of the earth do sufficiently demonstrate it to be full of invisible and underground fires," he wrote. "For wherever there is a volcano, there also is a conservatory or storehouse of fire under it; it is certain that where there is a chimney or smoke, there is fire. And these fires argue for deeper treasuries and storehouses of fire, in the very heart and inward bowels of the Earth."

Not only are there fires underground but great waters, which travel and pass through their own channels and estuaries, and according to Kircher, "the fire and water sweetly conspire together in mutual service." The tides, caused by the nitrous effluvia of the moon, push "an immense bulk of water" through "hidden and occult passages at the bottom of the Ocean" and thrust it "forcibly into the intimate bowels of the Earth." The resulting winds "excite and stir up" and otherwise feed the subterraneous fire like a huge bellows. The seas, which would stagnate and freeze without the heat, keep the fires going and also keep them from getting out of hand, preventing "unlimited eruptions," which would "soon turn all to ruins." Mountains, as suggested by Bernini's Fountain of the Four Rivers, are hollow, and function as huge reservoirs. Hot baths, hot springs, and fountains are produced where underground water passageways come near or interconnect with the fire channels.

More than once, Kircher compares the movement of the earth's water to the circulation of the blood as described by William Harvey. The water of the oceans follows "secret motions," known today as currents, leading up and around the globe toward the North Pole. Somewhere off the coast of Norway (the actual site of a major whirlpool system called the Moskenstraumen), he declares, is a giant maelstrom through which the water enters the earth, as if passing through a great drain. It runs through the earth's passageways, cooling it down, and providing it with elements and nutriments in particulate form before being eliminated through an opening at the South Pole. Sometimes the analogies refer more to the continuing process of the digestive system than to the cycling of blood, but no matter: "You see therefore the manner and way of the Circulation of Nature."

Fire and water work like light and shadow, consonance and dissonance, attraction and repulsion, playing their part in a more or less mystical totality that Kircher calls the "cosmos of the Earth" or the "geocosm." It is God's intent, he claims, for "both elements to be in perpetual motion, for admirable ends." Indeed, according to his scheme, the water provides the moisture and the fire provides the heat necessary to "fructify" the earth. The "fire in the belly of Nature," as he put it, is especially "necessary to the internal economy or constitution" of the earth, acting as a great furnace in which the "juices" of minerals, marbles, stones, and gems are melted and cooked, then mixed with the waters and cooled into their more familiar hardened forms.

Kircher's network of fires

KIRCHER'S NETWORK of oceans and fires had its idiosyncrasies, but his understanding of the way metals and various stones are made represented fairly common thinking of the period. It included the belief that subterranean processes of this kind, over time, eventually "ripened" base metals into gold. The idea that these processes might somehow be imitated and accelerated in the laboratory fueled much of the alchemical experimentation of the day. In the seventeenth century there was no clear distinction between alchemical practices and what might today be called legitimate chemistry. The *al* in *alchemy* is just an Arabic definite article ("the"), and *chemy* comes from *khymeia*, the Greek word for "fusion," which often referred to medicinal mix-

tures of organic substances, so *alchemy* really just meant "the chemistry." In *Underground World*, Kircher placed his lengthy discussion of
all related studies, from metallurgy to medicinal chemistry, under
that heading—alchemy.

He had read all the alchemical authors, including Zohara, Za-
dith, and Haled, but most notably Theophrastus Bombastus von
Hohenheim, otherwise known as Paracelsus, a Renaissance magus
whose own sources included the writings attributed to Hermes
Trismegistus and other ancient magicians. But in all of that research
Kircher had never found anything reliable about the legendary prime
material known as the philosopher's stone.

"The alchemists describe it as something wonderful and mysterious, which not only can cure the human body of all ills and keep it
healthy, but can change base metals into gold and silver," he wrote.
"They say it is a pure, unchanging, most simple metallic substance,
and that it is effective in infinitesimal amounts." But in a statement
all the more meaningful coming from someone not remembered for
disciplined thinking and restrained language, he proclaimed: "I came
to the conclusion that nothing was easier than to write in such a way,
putting down the first things that occurred to them, the most ridiculous fantasies of the human mind, in twisted words, solely to confuse
whoever tried to read them."

Kircher debunked a number of Paracelsus's claims through his
own experiments—following his instructions for making copper, for
example, and for converting various metals into quicksilver, with no
success. "Can one metal really be transmuted into another?" Kircher
asked. "In theory such a transmutation is possible, but in practice I
think it could only be accomplished with the help of devils or angels."

Moreover, he had no tolerance for the "frauds, deceits and other means by which the alchemists have pretended to make pure gold."

KIRCHER'S DISTASTE for what might be called improper alchemy didn't stop him from incorporating alchemical ideas into the rather amazing extended discussion of spontaneous generation that appears in *Underground World*, which is to say that he added alchemical language to the spiritual, Neoplatonic, Aristotelian, perhaps even atomic, and magnetic language he used to try to explicate the concept. In addition to repeating much of the material that had appeared in *Examination of the Plague*, and bolstering his argument with new experiments and new anecdotal evidence, he took his theorizing about *panspermia* to a new level: for all intents and purposes, universal sperm was the force of life. It was what Plato called "the world's seed," what Aristotle called "the moving power of all things," what Hermes called "the seed of Nature."

In the beginning, Kircher wrote, God created "a certain matter that we rightly call 'chaotic,'" out of which everything except the human soul was drawn, and in which is hidden this seminal power. "I say that a certain material *spiritus* was composed of the subtlest celestial breath, or from a portion of the elements, and that a certain spirituous salino-sulfuro-mercurial vapor, a universal seed of things, was created along with the elements by God as the origin of all things established in the world of corporeal entities." Where did Kircher get the idea that salt, sulfur, and mercury were principal? Paracelsus.

It's through this seminal power, quite simply, that nature propagates itself. Presumably, this power, both material and alive, is contained within the sperm of animals that procreate by mating. In other cases, the "salino-sulfuro-mercurial vapor" somehow individualizes itself as a kind of seed within whatever matrix of being it finds itself—animal, vegetable, or mineral. As Kircher described it, it consists of both a plastic power, which provides for physical form to take shape, and a magnetic power, about which he is rather vague.

Life can be engendered within the decay of a living being, Kircher claimed, because this vapor, or power, this "something" of the material soul, remains in the corpse, "not as a form but as spirituous corpuscles of this living being." Make no mistake, whether inside the decaying body or floating around seed-like, as it often does, it is in a highly degenerated and degraded state. That's why the living beings that grow from it are lower beings. But since these living things are not arising from utterly nonliving matter, he argued, it wasn't really proper to call this generation "spontaneous."

If it was difficult to articulate precisely how the life force functioned, it was nevertheless clearly at work when worms or maggots grew from rotting flesh—as everyone knew they did. It was at work when bees grew from the dung of bulls. It was at work when flies were engendered from the dead bodies of other flies. (Kircher said it helped to put them on a copper plate, sprinkle them with honey water, and expose them to the heat of ashes.) It was at work when live scorpions were born from the carcasses of dead scorpions. (You could assist with a little sunshine and sweet basil water.) And it was also at work when the mulberry tree produced the silkworm, which it did "on being impregnated with any chance animal."

————

AS A 1679 French write-up of *Underground World* put it, "It would take a whole journal to indicate everything remarkable in this work." The book included detailed charts of those "secret" oceanic currents, among the first ever published. Kircher's more or less correct explanation of how igneous rock is formed was also arguably the first in print. One modern scholar writes that Kircher "understood erosion," and his entries "on the quality and use of sand" and his "investigations into the tending of fields" had their practical use.

Underground World identified the location of the legendary lost island of Atlantis (something that modern science hasn't been able to accomplish) and the source of the Nile: it started in what is now South Africa as a number of little streams flowing down from the "Mountains of the Moon," then ran northward through "Guix," "Sorgola," and "Alata" and on into "Bagamidi" before reaching Ethiopia and Egypt.

Kircher offered a lengthy discussion on, for example, people who lived in caves (their societies and their economy), including the troglodytes he'd encountered in Malta. He reported on the remains of giants (also mainly cave dwellers) found in the ground, and went into detail on the kinds of lower animals who belong to the lower world, including dragons. "Since monstrous animals of this kind for the most part select their lairs and breeding-places in underground caverns, I have considered it proper to include them under the heading of subterraneous beasts," he explained. "I am aware that two kinds of this animal have been distinguished by authors, the one with, the other without wings. No one can or ought to doubt the latter kind of creature, unless perchance he dares to contradict Holy Scripture."

After all, "Daniel makes mention of the divine worship accorded to the dragon Bel by the Babylonians."

In short, *Underground World* covered almost every subject that might relate to the earthly sphere, as well as some that wouldn't seem to, such as the sun and "its special properties, by which it flows into the earthly world" and the "nature of the lunar body and its effects." These correspondences and influences were nothing new, but perhaps only Athanasius Kircher would choose to publish a series of moon maps in a book about the world below.

WHO WOULD READ such a book? People like John Locke, Benedict Spinoza, Christiaan Huygens, and Edmund Halley, to name a few. Later, people like Cotton Mather, Edgar Allan Poe, and Jules Verne. Henry Oldenburg, the secretary of the Royal Society, looked forward to its publication for several years, and in 1664 he wrote to Robert Boyle that the volume was soon expected at a London bookseller's. He subsequently paid more for it, fifty shillings, than for any other item in his very substantial library. Indeed a few years later, when Oldenburg made a list with the heading "Catalogue of my best books and what they cost me," three of Kircher's titles were included in the first five. It's hard to say exactly what he meant by "best," though, since *Underground World* presented certain problems.

Oldenburg wrote to Sir Robert Moray about one of the book's claims—that the moon caused the tides not by virtue of gravity but by virtue of a "Nitrous quality." Kircher had written that he could achieve the effect himself on a small scale. "Let it be experimented . . ." Oldenburg wrote, "whether Nitrous water, mixed with common salt, exposed in a basin to ye Beams of ye moon in a free open place and a cleere Moonshiny night, will boyle and bubble up."

Moray admired Kircher, though sometimes he found it difficult to champion him. (Among his comments about *Underground World*: "I do not deny it to be long.") He was perfectly willing to try the experiment, and studied the prepared bowl of water for the "large part of half an hour," but observed only a few small air bubbles. Robert Boyle set up a basin of water in the same way; his assistant stayed up two nights watching it, with no detectable result.

"'Tis an ill Omen, me thinks," wrote Oldenburg, "yt ye very first Experiment singled out by us of Kircher, failes, and yt 'tis likely, the next will doe so too."

To understate the case, Oldenburg wasn't the only one who had doubts about aspects of *Underground World*. In Florence, for example, a physician in the Medici court named Francesco Redi had several. When not composing poems such as *Bacchus in Tuscany*, his wine-celebrating dithyramb, Redi dissected animals and performed experiments in the laboratory of the Palazzo Pitti. After an education in Jesuit schools, he had taken to courtly life, and had recently begun wearing the kind of tall wig of ringlets made popular by Louis XIV of France. A leading member of the scientific society established by Grand Duke Ferdinand II, and head of the ducal pharmacy, he was a proponent of the new philosophy and the experimental method. Redi wasn't completely modern; as a physician, he was just as apt to prescribe plant-derived purges and donkey's milk as the next early modern doctor. But Kircher's section on spontaneous generation struck him as dubious, and he decided to look into the age-old idea for himself.

In his *Esperienze Intorno alla Generazione degl' Insetti* (*Experiments on the Generation of Insects*), published three years later, Redi described a series of close observations involving a variety of rotting animal tissue: "a large pigeon," "a sheep's heart," a "large piece of horseflesh," "some skinned river frogs." As expected, maggots appeared. But there was something else. "Almost always I saw that the decaying flesh and the fissures in the boxes where it lay were covered not alone with worms, but with the eggs from which . . . the worms were hatched." Also a lot of flies were hovering around.

Redi then conducted one of the very first controlled experiments ever documented: "I put a snake, some fish, some eels of Arno, and a slice of milk-fed veal in four large, wide-mouthed flasks; having well closed and sealed them, I then filled the same number of flasks in the same way, only leaving these open." Maggots soon began to appear on the flesh in the open containers, but not in the closed ones.

He experimented with a number of variations, sometimes covering the flasks with "a fine Naples veil," but no maggots or worms or anything at all ever appeared inside the covered containers. He also dutifully followed Kircher's instructions for breeding bees in the dung of an ox ("I don't know whether that estimable author had ever carefully made this experiment, but when I made it . . . I observed no generation of any kind."); for breeding scorpions in dead scorpions ("I risked a second and third experiment, only to be disappointed and to wait in vain for the desired young scorpions . . ."); and for breeding flies in dead flies ("I believe . . . that the aforesaid honeywater only serves to attract the living flies to breed in the corpses of their comrades and to drop eggs therein."). He never had any luck.

With no insects appearing in closed flasks, Redi was convinced that "no animal of any kind is ever bred in dead flesh unless there be a previous egg deposit."

Redi referred to Kircher somewhat patronizingly as "a man of worthy esteem," but it had never occurred to Kircher to regulate his experiments this way—to put a lid, as it were, on his container. That was really all it took to disprove what Redi called "the dictum of ancients and moderns." It was as if Redi were making a strangely elegant threat to all long-held assumptions everywhere, while Kircher was trying to reaffirm a conception of the world he sensed was slipping away.

Kircher, who turned sixty-three the year *Underground World* came out, wasn't willing to relinquish a thing. By 1668, when Redi's manuscript for *Experiments on the Generation of Insects* was ready for publication, Kircher had already published *five* more books, on everything from the mystical significance of numbers to the history of the mountain shrine at Mentorella he was restoring (in his free time).

17

Fombom

The experimental circumstances in which Kircher allowed a dog to be bitten by a venomous snake came about even before work on *Underground World* was finished.

The snake came in a wooden crate along with other vipers destined for the pharmacy of the Collegio Romano, where, in accordance with centuries-old procedures, their meat would serve as the chief ingredient in a new batch of theriaca, the fermented concoction believed to heal or prevent all kinds of afflictions, especially the bites of snakes. (Theriaca was supposed to work by sympathetic means. Those means might also be described, in Kircherian terms, as magnetic. In modern terms, they might be called homeopathic.) Some recipes called for more than sixty ingredients, including opium, rhubarb, nutmeg, turpentine, St. John's wort, and the yellow secretion from the castor sacs of beavers. Apothecaries liked to age their theriaca on the shelf for a number of years.

The arrival of the vipers represented an opportunity for Kircher to test a medical curiosity. A Bavarian-born Jesuit missionary named Heinrich Roth had recently visited Rome after years at Agra, site of the recently built Taj Mahal, and on the island of Salsete, near Bombay. He'd presented Kircher with three so-called serpent stones, or snake stones. These lightweight items—about the size of a small coin and inconsistently described as reddish, green, or white, with brown or blue around the edges, or black—were said to have been cut from the heads of cobra snakes found in India, China, and Southeast Asia, and to serve as the only antidote to the cobra's bite. (Cobra is short for *cobra de capello*, Portuguese for "snake with a hood.") When applied to a snakebite, the stone supposedly adhered to the skin, drew out the venom, and fell off when saturated with poison. These little stones or bones were beginning to show up in courts and salons all over Europe. Jesuits presented specimens to the Holy Roman Emperor, and Franciscan missionaries offered a number of samples to the Medici court. The repository of the Royal Society had a stone from "Java Major," now called Sumatra. There were reports from Jesuits that they worked.

As Kircher described the trial, he gathered a "multitude of Fathers and other curious men" to observe, and then arranged for the hapless dog and one of the newly arrived vipers, known for the depth and tenacity of their bite, to meet. The viper struck. The snake stone was applied.

"When this stone was placed on the dog's snake bite, it stuck to the wound so that one could scarcely pull it away, remaining fixed to the wound for a long time," Kircher reported. "Finally, having drained all the poison, it fell away by itself, like a leech saturated with blood. The dog was free from the poison, and although feverish for a while, was restored to his former health after about a day."

Kircher included the account in his very popular *China Illustrated*, published in 1667. But there was a problem with this experiment:

Kircher may have only *claimed* to conduct it. His rendering of the event is a little too similar to the account of another Jesuit, published in 1656. (Kircher: "I wouldn't believe this, unless I had done an experiment with a dog who had been bitten by a viper." Father Michał Boym: "I wouldn't have believed it myself unless I had performed an experiment on a dog.")

Perhaps because the report came from a fellow Jesuit, there was no real need to conduct that particular test, only to say that he had. Maybe he was just replicating the experiment as meticulously as possible. In whatever trials he did make, he very probably observed a tendency on the part of the snake stone to stick to an open wound. But rather than look into the porosity or absorbency of the unusually lightweight material (if not actually dried organic material, it may have been pumice stone), he saw what he was searching for, and attributed the snake stone's apparent drawing power to magnetism.

To Kircher, this "miraculous magnet of poisons" was a new piece of evidence in a larger argument that deserved more than a mention in his book on China. A public reaffirmation of his magnetic philosophy was in order. As he wrote in the resulting *Magnetic Kingdom of Nature*, also published in 1667, "I think that the immutable force of nature implanted in particular things which does not proceed from manifest or elemental qualities ought to be called magnetism."

Whenever invisible forces were at work, he claimed, magnetism was at work. Magnetism is "the same thing which is called occult by some, or the instrument of divine potency by the Hebrews, or the hidden form operating in all things by others; some call it the sympathetic and antipathetic quality. But I believe it should be called magnetism, since in truth all energy existing in things of this kind works according to the analogy of the lodestone, that is, attraction and repulsion."

A modern academic writes that Kircher "ignited wide publicity about this wonder and its medical powers" and became "the leading advocate for the efficacy of the new therapy." Robert Boyle would conduct trials. Francesco Redi would, too.

ONE WAY OR ANOTHER, Kircher's notoriety grew, and his publishers in Amsterdam had no qualms about trying to capitalize on it. *China Illustrated*, for example, which was in fact copiously illustrated, was meant to feed European curiosity for more reliable information about the culture behind gunpowder, paper, porcelain (or "china"), and the parasol. Was it true, for instance, that Chinese physicians could cure sickness without bloodletting? Readers looked to someone of Kircher's stature to augment their fairly paltry knowledge of China with his patented blend of virtually everything that was interesting and important to understand.

First among these things, as he explained it, was that the lands of the East were originally settled and populated by Egyptians after the Great Flood. Kircher believed the Egyptian roots of Chinese culture could clearly be seen in its caste system and in its penchant for "mystic temples," but most obviously in the written language, the hieroglyphics, of the Chinese. The Chinese also honored the teachings of Hermes Trismegistus, though they called him Confucius. And they worshipped the Egyptian goddess Isis, only they called her Pussa. (In Japan, Kircher claimed, they worshipped a male version called Fombom.)

The goddess Pussa on a lotus, from China Illustrated

Drawn primarily from the reports of missionaries who had been traveling to the East since the 1570s, *China Illustrated* didn't limit itself to China. Hence the section on the use of the snake stone, which Kircher attributed primarily to the Brahmans in India. But there was no shortage of fascinating things to note about the Chinese empire itself, which stretched from the tropics to the "cold and frozen northern zones." "It is so large that you can't find a more populous nation anywhere on earth," he reported, estimating the figure to be around two hundred million, "not counting the royal ministers, eunuchs, women, and slaves."

Interesting creatures of the East included the "fast cow" (the rhinoceros), the "marine horse" (the hippopotamus), and various species of ape, which Kircher noted were very much like humans: "Except for the foulness of their bottoms you would scarcely believe they were animals."

There were many beneficial roots and herbs and aromatic oils and woods, as well as natural curiosities. There was a river in Quandong that was supposed to turn blue in autumn and a rose that was supposed to change color twice a day. There was also a drink called tea, which he said was "gradually being introduced in Europe." Consumption of this beverage is the "main reason there is no gout or stones in China." It "keeps the oppression of sleep away from those who want to study." It "is also used for relieving a hangover, and one soon can safely drink again."

As with many of his books, this one got a lot wrong *and* had a significant influence on its readers. One of the first widely available Western books on the subject, it was so well received that a second Latin edition was published within a year. Jansson and Weyerstraet prepared French and Dutch translations too, and further editions and excerpts were printed in Rome, Antwerp, and London, where the book at least contributed to the English fascination with Chinese language, architecture, and consumption of tea. (A London coffeehouse in Exchange Alley named Garraway's was one of the few to serve tea at the time the book appeared. The British East India Company began to import tea in 1668, a year after *China Illustrated* was published.) Historian Charles D. Van Tuyl, who published an English translation of the book in 1987, says that it "was probably the single most important written source for shaping the Western understanding of China and its neighbors."

Kircher's publishers in Amsterdam had already produced or were planning new editions of *Universal Music-making*, *Underground World*, and *The Great Art of Light and Shadow*. Translations and editions of other works had been printed in Leipzig, Würzburg, and Cologne. Copies of his books were in the possession of the Margrave of Brandenburg, the Duke of Saxony, the Duke of Waldstein, and the Duke of Schleswig. They were sold by book merchants in Vienna, London, Naples, Venice, Paris, and Madrid. And they were shipped, carted, and generally lugged to Jesuit colleges and missions all over the world. One French missionary took twenty-four copies of *Universal Music-making* and twelve copies of the four-volume *Egyptian Oedipus* with him from Lisbon to Peking.

Kircher "had a global reputation," says a twenty-first-century scholar, "that was virtually unsurpassed by any early modern author." And with that fame sometimes came obsessive admirers. When a certain Criollo priest in New Spain named Alejandro Favián first came into contact with *Universal Music-making*, for instance, he was transformed by the experience. "Truly without exaggeration," he wrote, "I say nothing better has ever happened to me in my life." He idled away the hours not only reading Kircher's books and writing him letters but staring at his portrait, which he'd decorated with feathers and gold. Favián even built a museum in which he hoped to display musical machines and optical devices. In imitation of his master he wrote a massive book of his own that he asked Kircher to help him publish. Titled *Universal Ecstatic Tautology*, it sounded more like a send-up of Kircher's work than homage. But the five-volume, three-thousand-page tome was evidently an earnest attempt to encompass all things.

IT WOULD HAVE COME as no surprise to Favián or other infatuated readers that back in Rome the person they assumed was the greatest scholar of his time might collaborate again with the greatest artist of his time on another one of Rome's popular landmarks.

In September 1665, the Dominicans of Santa Maria sopra Minerva (St. Mary Above Minerva)—in whose convent Galileo was tried—were digging the foundations of a wall in their garden, when they found the remains of an obelisk. The site, in the shadow of the Pantheon and a few hundred feet from the Collegio Romano, was known to have been an ancient temple of Isis (Minerva to the Romans). Kircher wrote that his friend the pope sent for him right away: "Upon learning of the matter, without hesitation, His Own Blessedness summoned me to himself and entrusted to me the charge of examining the situation." It was the pope's will "that the ruined obelisk be revealed to the public light as quickly as possible in order to hasten the interpretation of the mysteries which are contained in it."

Kircher was unable to stay in Rome, because of the "approaching solemnity" of an "apostolic mission" at the shrine of Mentorella. So he delegated the task of copying down the markings on the obelisk to a student named Gioseffo Petrucci—"my assistant in the studies of Egyptian antiquities"—ordering Petrucci to send the scheme to him as soon as possible. But the young man, described elsewhere as "secular, originally from Lombardy," could copy only three of the monument's four sides: "The fourth side was lost on account of the difficulty of rolling the obelisk." (At about eighteen feet, this obelisk was one of the smallest found in Rome, but eighteen feet of solid granite weighs a figurative if not a literal ton.)

But Kircher wasn't put off by this limitation: "I (let there be praise and honor and glory to God), after completing a most precise scrutiny of the obelisk, grasped the entire series of mysteries hidden beneath it in such a way that not even that fourth side, which had been omitted from the delineation because it was hidden, escaped my comprehension." It would be risky to share his delineation of all four sides, but he was sure of himself. "Rather bold perhaps in my confidence, although in no way insecure since all ambiguity had been put aside, I sent to Petrucci in Rome the yet uncovered fourth side's scheme."

As Kircher told it, Petrucci was "thunderstruck." He called together the Dominican fathers as well as some of "the more experienced literati" of Rome. "They in turn marveled at my boldness," Kircher claimed, "and perhaps my lack of temerity, but several decreed that the truth of the matter must be determined by the original on the obelisk itself."

After the obelisk had finally been rolled, they compared Kircher's scheme with the newly revealed side. "And when they had discovered that soundly and without error all of my markings were composed as on the original," he recalled, "they were utterly stupefied, those same men who were formerly mocking my interpretations as merely pure conjecture."

This left "certain individuals saying that this knowledge had been inspired by the power of God, while several, not without calumny, even asserted that the knowledge had been acquired by some illicit pact with a demon. Some, finally, judged that this type of knowledge, attained by many years of study, was able to be acquired by the strength of a singular intellect."

The pope was of the last opinion, and even asked Kircher for private instruction in hieroglyphics. He wanted to understand the sacred meaning of the newly discovered monument that, in a sort of papal tradition, would be re-erected and "inscribed with the glorious title of his own name." As Kircher interpreted the obelisk, the sacred meaning had to do with the way the

> supreme spirit and archetype infuses its virtue and gifts in the soul of the sidereal world, that is the solar spirit subject to it, from whence comes the vital motion in the material or elemental world, and abundance of all things and variety of species arises.

Again it seemed that the Egyptians believed in the same kind of panspermatic solar abundance that Kircher espoused. It "flows ceaselessly," according to his translation, because it is "drawn by some marvelous sympathy" that sounds an awful lot like magnetic attraction.

Together with Bernini, who had recently returned from the court of Louis XIV in France, Kircher and the pope made plans for the obelisk to go in the square in front of Santa Maria sopra Minerva, near where it had been found. Bernini's idea to put the monument on the back of a marble elephant alluded to a long tradition in which elephants were associated with intelligence as well as physical strength. But by the time the obelisk was dedicated in July of 1667, ever feeble Chigi had died, and it became a memorial to him. The Latin inscription on the base was written by Kircher:

> Everyone who sees the images carved on the obelisk by the wise Egyptians and carried by the elephant, the strongest of

beasts, should understand: a robust mind is required to sustain solid wisdom.

The pope never had a robust body, but his mind had sustained enough wisdom to inspire one of Rome's favorite sculptures. Romans endearingly called the elephant *il porcino* ("the little pig") until a few hundred years ago, when that morphed into *il pulcino* ("the little chick"). Kircher said it was a privilege to "bear the honor of observing him by erecting the Alexandrian obelisk." But it was Kircher, after all, who revealed the divine wisdom, such as it was, to Chigi; it was Kircher who had worked so hard to penetrate the hidden mysteries of the hieroglyphics in the first place. And so, in helping erect this monument to Chigi's robust mind, he was also erecting one to his own.

KIRCHER HAD ONCE even dared to dream, literally, that *he* might be made pope. According to a report published by Schott, he became "stricken with grave and perilous disease," and after prescribing "soporific medicines" to himself, fell into a deep sleep. "He dreamt that he had been elected Supreme Pontiff and that he had garnered the legations and congratulations of the Christian Princes and the applauses of all nations, a thing which suffused him with boundless joy." The dream somehow renewed him.

Now, in Rome, many people were secretly relieved by the death of the "morbidly austere" Chigi, and they celebrated the announcement of the man who had been elected pope in reality, rather than in his dreams. The new pope, Giulio Rospigliosi, took the name

Clement IX. He wrote comic opera librettos and enjoyed evenings out. To the job of secretary of state, he appointed the cardinal who was said to be Queen Christina's lover. Christina, who had been on tours of Paris and Hamburg, returned to Rome and accepted a stipend. She helped Clement establish the first public opera house in the city, and helped persuade him to prohibit the racing of Jews during Carnival. (The prostitute races continued.)

In addition to putting on plays of the *sporchissime* (very dirty) sort, and hosting the best harpsichordists and castrati at her palazzo, Christina took up archaeology, about which Kircher claimed some expertise, and alchemy, which Kircher had disparaged. She also held regular meetings of a literary and intellectual salon, which as a first order of business had banished overdone language and embellishment, the kind that Kircher had embraced.

Kircher knew that some people thought his hieroglyphic interpretations were "pure conjecture." The story of how he delineated the fourth side of the obelisk may have enhanced his reputation among those who already idolized him, but it didn't "stupefy" quite as many of his detractors as he may have hoped. Those who went to see the obelisk must have noticed that the markings on each side were actually quite similar; it might not be too difficult to guess the fourth side, especially after almost thirty years of hieroglyphic study. To those who were dubious, Kircher was self-aggrandizing, and his self-proclaimed mastery of the hieroglyphics set him up for criticism and ridicule.

Stories about Kircher began to travel within the social circles of Rome and the salons of other cities. Christina wrote in her memoirs, for example, about an incident involving a philologist named

Andreas Müller. Müller concocted an utterly unintelligible manuscript, then sent it to Kircher with a note saying it had come from Egypt, and asking for a translation. Kircher apparently produced one right away.

Once the word got out about these sorts of practical jokes, they were passed along with such frequency, and set down in so many different forms, that it became impossible to say for sure which tricks had actually been played. Was it "some mischievous youths of Rome" or a single "wicked wag" who had an old stone engraved with nonsense and then buried it one night at a site where workers were digging? When Kircher was called to interpret this stone, did he say that he needed time to try to discover the meaning, or did he begin, as one account had it, "to leap and dance for joy—and to give a beautiful interpretation of the circles, the crosses, and all the other meaningless signs"?

One anecdote comes by way of the American author, critic, and newspaperman H. L. Mencken. In 1937, Mencken published *The Charlatanry of the Learned*, an English translation of a book originally printed in the eighteenth century by an ancestor named Johann Burkhard Mencken. As the story goes, Kircher was given a piece of silk paper containing some intriguing, odd-looking characters. After he spent a number of days trying to decipher it, he was finally taken to a mirror and shown that it was merely Latin written in reverse: *Noli vana sectari et tempus perdere nugis nihil proficientibus*, the message read. "Do not seek vain things, or waste time on unprofitable trifles."

18

Everything

Recall that for the Jesuits, the path toward Christ was predicated on an effort to achieve humility. It's unclear how well or how often Kircher took a good look at his apparent lack in that regard. But given that hypocrisy is almost requisitely present in human beings, and common among religious, political, and philosophical practices, he surely wasn't the only vain or self-interested member of the Society of Jesus.

As the satirizing sermon by a monk from another order in Rome went, the Jesuits "are the best Men that Live on the Earth. They are as Modest as Angels. They never open their eyes to cast a Look upon the Ladies at Church. They are such great Lovers of Restraint, that you never see them in the Streets. They are so in Love with Poverty, that they Despise and trample upon all the Riches in the World. They never come near Dying Persons or Widows, to importune them to be Remember'd in their last Wills. . . . They never go among Courts, or mind State Affairs."

If the question among Jesuit authorities was whether Kircher was sometimes too concerned with advancing his own name, the answer may have been that even so, he had also advanced the interests of both the Jesuits and the Church.

It's said that as he got older, Kircher spent long periods of time in contemplation at the shrine of Mentorella. "Those letters you have sent to me," he wrote a friend in his later years, "I have read with equal affection of the spirit, not in Rome, but established in the vast solitude of the Eustachian or Vulturellian mountain, to which I am accustomed to take myself during the autumn holidays, in order that, free from every worldly noise and with cares of studies somewhat set aside, I might be able to conduct the business of health with God, a business, if anything, of exceedingly great importance in the world."

Kircher certainly threw himself into the restoration of the shrine there with the same energy he put into his more prominent projects. To help pay for it, he "procured aid" from Leopold, the Holy Roman Emperor, the Duke of Bavaria, the archbishop of Prague, and "the most excellent" king of Naples.

"I firstly fitted the Church with every manner of magnificent preparation, both pictures and tapestries," he later wrote of the renovation. "I restored the ruined altars with concrete and . . . those ornaments which would be necessary . . . for celebrating mass." To the church, Kircher "joined a structure conspicuous for its thirteen vaults and complete symmetry." And since there was no way of reaching the "very lofty peak" where Saint Eustace was supposed to have seen the stag, "we built to the crag's peak a staircase constructed with huge rocks . . . and atop the crag I built a chapel consecrated to divine Eustachius." Then, "since all these arrangements would be in vain were there no people who would visit the place for the sake of their devotion," he established an apostolic mission. "Yearly on the

festival of the Archangel Saint Michael with the solemn promulga-
tion of complaisance nearly many thousands of people of each sex
flocked to participate in the sacraments."

Even these somewhat oversize acts of devotion smack of self-
regard and a sense of personal achievement. But if Kircher ever had
any scruples about his motivations, he could turn to the guidance
that Ignatius of Loyola had provided on the subject. "A person may
wish to say or do something consistent with the Church, something
that promotes the Glory of God," Ignatius wrote. "In those circumstances, if a thought, or rather a temptation, comes from without not
to say or do that thing, proposing specious arguments about vainglory or something else, then such a person ought to raise the understanding to our Creator and Lord." Then, if "one sees that the
proposed action is for God's service, or at least not against it, one
must act in a way opposed . . . to the temptation."

In Kircher's case, accusations of vainglory didn't stand a chance.
There were still things to be done for the greater glory of God, and
the next thing to do was, in a word, everything. Kircher set his sights
on a new method for understanding all that could be known.

OTHERS IN KIRCHER'S LIFETIME had attempted a new, comprehensive approach to knowledge—Francis Bacon and René Descartes,
to name two. But they hadn't produced anything quite like this. *Ars
Magna Sciendi (The Great Art of Knowing)* was one of the most
strangely beautiful books Kircher ever produced, full of symbols and
symbolic formulations. It *looked* as if it might somehow get down
to the essential relationship of things, might establish "an order between all thoughts which can enter the human mind," as Descartes

had said a universal language must. But unlike Bacon and Descartes, Kircher wasn't interested in tearing down or throwing away all previous learning in order to create his method for determining the truth. And it wasn't skepticism of received doctrines that drove his thinking about the possibility of the project in the first place. It was the wisdom of the ancient philosopher Plato, who said that "nothing is more beautiful than to know all things." Kircher believed it was possible to experience this kind of beauty, or rather it was possible to teach others how to, since presumably he knew. The mechanism for his universal path of inquiry that could be applied to every field? The combinatory method of Ramon Llull that he'd employed in his music composition system and his mathematical organs.

This great art of knowing was, as Kircher described it, "the art of arts," "the workshop of the sciences, the fertilizing seedbed of the minds, the key to all human cognition, by which a most complete approach lies open to the understanding of all things which pertain under the notion of the intellect."

The idea was that the universe was arranged according to certain principles—or that certain principles were inherent in its arrangement. These concepts made for a kind of matrix or architecture by which all nature is engendered. In order to "understand and love God," as Llull had it, one had to organize the mind in the same way.

Basing his work on the Llullist system, Kircher started out with nine essential principles or attributes of God (a trinity of trinities): Goodness; Magnitude; Duration; Strength; Wisdom; Will; Power; Truth; and Glory. There were nine universal subjects: God; Angels; Heaven; Elements; Man; Animals; Vegetables; Minerals; and Numbers. And nine "respective" principles, or principles of relationship: Difference; Agreement; Opposition; Beginning; Middle; End; Majority; Equality; Minority.

Each of these was given an icon that could transcend the barriers of language. According to Kircher, with the addition of nine interrogatives (for putting statements to test via syllogism), these twenty-seven principles can produce a total of 371,993,326,789,901,217,467,999, 148,150,835,200,000,000 concepts for consideration. This universal method could be applied within any discipline, he explained, and the icons supposedly made the system accessible to people of all languages. He may have imagined it as a missionary tool that could be used by Jesuits around the world. But a great many assumptions have to be made to come up with these principles and subjects in the first place, or to adopt them wholesale from Llull, as in effect Kircher did. In order to get anywhere with them, one really already had to have a philosophy of knowledge and truth in place.

As Umberto Eco has described Llull's original system, the combinations "do not generate fresh questions, nor do they furnish new proofs. They generate instead standard answers to an already established set of questions. . . . It is, in reality, a sort of dialectical thesaurus, a mnemonic aid for finding out an array of standard arguments able to demonstrate an already known truth."

Still, the notion of so many combinations might inspire wonder. The graphs in the book, which allowed readers to visualize the possibilities, seemed almost to shimmer. Perhaps Kircher believed that readers might begin to "know" the sublimity of the universe by contemplating these alone. This was the art, not the science, of knowing, after all. It didn't help determine truth so much as help see the "truth"—that all is one, that "the least is in the highest, the inter-

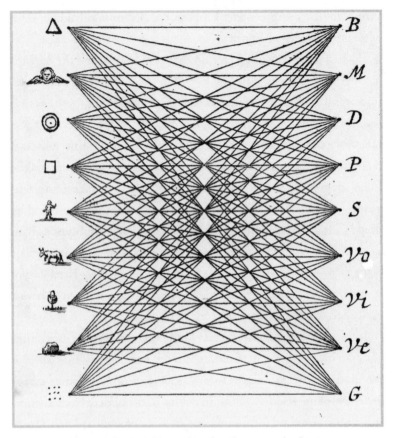

Essential principles combined with universal subjects

mediate is in both the lowest and highest; that the highest is in both the mediate and the lowest; and, in a word, that everything is in everything, each in its own way."

THE *PHILOSOPHICAL TRANSACTIONS OF THE ROYAL SOCIETY* provided a report on this "Voluminous Work," which, it said, proposed "to enable men to discourse and dispute, innumerable ways, of everything proposed, and to acquire a summary and general knowledge of all things."

"Of what Use this Doctrine may be for the attainment of knowledge with more ease and advantage," said the review, "the sagacious reader may judge."

But there was an intensely intellectual young man named Gottfried Wilhelm Leibniz who was very interested in *The Great Art of Knowing*, and wrote to Kircher to tell him so. Leibniz is most famous for inventing calculus independently of Isaac Newton, and for developing the binary system, on which computer processing relies. In May of 1670, when he wrote his long and fawning letter, he was a self-conscious and skinny twenty-three-year-old, a young legal advisor in the court of the Prince-Elector of Mainz, where Kircher had served almost fifty years before. He'd entered the University of Leipzig at only fourteen to study law and philosophy. His first job after school was as a professional alchemist in Nuremberg—he had impressed his prospective employer with a dense alchemical treatise he'd entirely made up. Or so the story went; Leibniz himself may have made *that* story up later in life in order to play down his early and very serious interest in alchemy.

By the time he wrote Kircher, he'd already mapped out a plan to reform the practice of law in the Holy Roman Empire, crafted two essays on physics, started work on a calculating machine, and begun devising a political scheme to redirect French ambitions away from German lands and toward Egypt. (More than a century later this particular piece of bait was taken by Napoleon.) But he had even larger, more idealistic, and overarching ideas in mind, and it wouldn't be a stretch to say that many were Kircherian in nature. As a modern historian has said about Leibniz's subsequent intellectual career, "Virtually every major scientific, linguistic, and historical project on which he embarked had been directly inspired by reading Kircher's works." Even Leibniz's idea to use wind power to drain the water from the silver mines of the Harz Mountains seems to have come from Kircher's proposal, published in *Underground World*, for ventilating mines with giant weather vanes.

As a boy, Leibniz came across Kircher's books in the process of devouring the volumes in the library of his late father, a professor of moral philosophy. Having grown up in German lands laid waste during the Thirty Years War, he shared Kircher's general desire for unity and for synthesis—of not only the political but also the religious, intellectual, and philosophical sort. He was a believer in "the elegance and harmony of the world," and in the notion of a single, universal Christian church. This non-practicing Lutheran had even begun mounting a defense of Catholic tenets through a series of essays called *Catholic Demonstrations*, with the goal of proving heretics, skeptics, and certain new thinkers wrong. His defense of transubstantiation, for example, took into account the "philosophy of the moderns," and in his treatise on the physics of motion, he employed mathematics to make rigorous, almost legalistic arguments for the mind-like qualities of matter.

Like Kircher, Leibniz applied himself to the pursuit of a universal language and to a universal philosophical approach that could be brought to bear on every possible question to produce universally accepted answers. Also like Kircher, Leibniz was an opportunistic courtier and a flatterer to the highest degree. The 1670 letter wasn't his first attempt to get Kircher's attention. Four years before, at the age of nineteen, while also working toward his doctorate in law, Leibniz had written his own dissertation on the Llullian combinatorial arts, and sent Kircher a copy, but received no response. Now, after reading *The Great Art of Knowing*, he wrote again. Calling Kircher a "GREAT MAN," the "greatest man," "an incomparable man," and a "man worthy of immortality" (a play on the meaning of *Athanasius*), Leibniz praised his latest work:

> I have come by chance in a most happy year upon your work about the great art of knowing or combination; I have drunk it deeply, I have read it, I have wished upon it most avidly, and I have not put it down until it was completed; what more can I say? You filled me entirely with admiration and love for you . . .

In fact, young Leibniz told Kircher, "no mortal to this day has penetrated so deeply as you into the secret art of combination." And he was one to know, since he himself "strenuously, nearly since my childhood," had engaged in many "cogitations on this matter."

What Leibniz wanted, it seems, was to develop a correspondence with Kircher to exchange ideas, and also to gain a highly visible benefactor or mentor. Referring vaguely to "new ways of dealing in syllogisms," and to the "universal, advantageous" ways of bringing figures into the process of "true proclamations," Leibniz wrote that he had often wished for "one hour" of conversation with Kircher in which to share his ideas—"grand things indeed but nevertheless with an eye to the public good." He offered his services as a "constant, energetic public announcer" on Kircher's behalf, suggesting he'd be happy even to hear from one of Kircher's "substitutes."

After continuing this way through two pages of minute ink scratchings, he was about to sign off, "except that one thing is left that is worthy of asking." Leibniz was referring to "book 3, part 5, chapter 4, problem 1," of Kircher's *The Magnet*. "You report," he wrote, "that in the Arabian Market of Marseilles, you found a certain miraculous material, which even when covered turns itself towards the sun, of what sort no Heliotrope has been known to do."

Of course: the special sunflower seeds or sunflower-like material that Kircher had been intentionally vague about, the claim that also caught Descartes's attention. Two of the most brilliant minds of the seventeenth century had stopped themselves over the same section in a nine-hundred-plus-page book.

Leibniz was a bit confused, or pretended to be. "Concerning this thing," he wrote, "if it is allowed that I ask, what is the name of the material, whether it is a mineral or something of some king, and from which part of Arabia was the most powerful Market, or from what part did the material originate?"

KIRCHER WAS RATHER condescending in his response to the young courtier from Mainz. He may have had other things on his mind. By debunking alchemy in *Underground World*, for example, he had provoked a firestorm of criticism. In the words of one nineteenth-century writer, "All the alchymists were in arms immediately to refute this formidable antagonist." Johann Glauber, a prominent alchemist-pharmacist, and Johann Zwefler, a prominent alchemist-physician, suggested that Kircher's hostility toward them was a form of sour grapes; he had tried but failed to master the al-chemical art as they had. One professor from Padua named Bonvicini went so far as to claim that he was actually in possession of the philosopher's stone. As Kircher reported, Bonvicini was then summoned by the Senate of Venice "to assist them with his gold-making art in paying the expenses of the war against the Turks." Unable to make good on his claims, "his mind collapsed and he fell into a melancholia, of which he died eight days later."

One critic, a "true pupil of this art" going by the name Salomone de Blauenstein, published a tract, *In Defense of the Philosophers' Stone, In Opposition to the Anti-Alchemy Mundus Subterraneus of the Jesuit Father Kircher in which not only are his arguments calculated against alchemy refuted, but even the skill itself is made manifest to intelligent men insofar as it is possible.* His evidence included a story about a Pole named Sendigovius who changed quicksilver into gold in the presence of the Holy Roman Emperor sometime around the turn of the seventeenth century—it had to be true, since the emperor had commemorated the transmutation with a plaque. He also engaged in the increasingly popular pastime of pointing out Kircher's errors. Kircher was under the incorrect impression, for example, that Arnold of Villanova, the thirteenth-century physician to whom many alchemical tracts and the discovery of carbon monoxide are attributed, was actually two different people: Arnold and Villanova.

"When he read how I had very clearly revealed all the vanities and impostures of the alchemists, as well as the chimera of the Philosopher's Stone . . ." Kircher explained in a letter, "he was thrown into a rage." To Kircher, this "Blauenstein" person was not only "an imposter" but "a lying clown and a defrocked monk."

PART THREE

19

Not As It Was

As 1671 began, the aging Kircher had reason to feel optimistic. His publishers in Amsterdam put out a beautifully illustrated edition of *Latium*, his study of the area around Rome. It read less like a work of history than a tour through the surrounding countryside by way of ancient literature. According to Kircher, Noah's family had indeed settled in the region, making it "the primeval seat and colony of the earliest mortals." Moreover, he said, Noah's traits and qualities served as the inspiration for the gods of Roman and Greek mythology. The book was really an attempt to reconcile old texts, Roman mythology, and the Bible with the actual geography and archaeological evidence of the place. The maps and engravings depicted the region "not as it was," as he strangely put it in one case, "but as it could and must have been."

Those who had devoted much of their lives to studying the topography and ruins of the region unfortunately found this otherwise handsome book to be, in the words of a modern writer, "flawed beyond belief." One of these scholars published an entire catalog of the errors to be found in *Latium*, wondering in print whether Kircher had ever actually been to some of the sites he wrote about.

More embarrassment came from the laboratory of Francesco Redi in Florence. After publishing *Experiments on the Generation of Insects*, in which he disproved Kircher's claims about spontaneous generation, Redi found himself reading Kircher's accounts of the healing power of the snake stone in *China Illustrated* and *The Magnetic Kingdom of Nature*. It happened that Redi had taken up the topic of snake stones, too. Plenty of them had been presented to the grand duke by travelers from abroad. Redi was head of the Medici pharmacy, which produced its own vintages of viper-rich theriaca, so he also had plenty of snakes on hand. Over several years he'd conducted five sets of experiments in front of witnesses, using a wide selection of snake stones on a range of bird species, including guinea hens, rock pigeons, and barnyard chickens. In fact, Redi employed a total of two hundred fifty vipers in these and other experiments on snake venom and toxicology. The porous snake stone did have a tendency to stick to the skin, but except in a few of what Redi called "freak" instances, all the birds and animals that had been bitten by vipers died.

In 1671, Redi published his conclusions, that snake stones did not heal victims of poisoning, in a public letter addressed directly to Kircher: *Experiments on various natural things and in particular on those that have been brought from India, carried out by Francesco Redi and described in a Letter to Father Athanasius Kircher of the Company of Jesus*. As Redi later explained: "The principal point of this letter was for me the experiments which I conducted with this stone, which notwithstanding the witnessing of so many authors, has always, always proved itself to me in all trials most useless and of no value."

At the same time, an Englishman, Sir Samuel Morland, had been testing an amplifying device he called the *tuba Stentoro-phonica*—also known as the "speaking trumpet." In the twenty-first century this device is known as the megaphone. The king's master of mechanics, Morland had fabricated initial prototypes out of glass and brass but settled on copper as his material of choice. Charles II himself, along with Prince Rupert and other members of the English nobility, participated in a trial in St. James Park. "Standing at the end of the Mall near Old Spring-Garden," according to Morland's account, they heard Morland "word for word" from the other end, about eight hundred fifty yards away.

In another trial, statements made through a sixteen-foot trumpet at Cuckold's Point could be heard in a rowboat about a mile and a half down the Thames. At the king's direction, further experiments were conducted off the ramparts at Deal Castle up on the English coast: with the wind blowing from the shore, words spoken through a twenty-one-foot horn were understood as far as three miles out in the channel. The king, persuaded of the many uses of the speaking trumpet—among them commands to whole fleets, commands to whole armies, commands to hundreds of workers, messages of relief to citizens of besieged cities and towns, and messages of intimidation and hostility to the citizens of besieged cities and towns—ordered that a number of them be made. The speaking trumpet caused a kind of sensation. A smaller version of the horn could be purchased together with Morland's little volume on the subject in the shop of Moses Pitt, a prominent London bookseller. It wasn't long before the trumpets were being sold in various sizes and dimensions around Europe.

Morland accounted for the effect in (erroneous) terms reminiscent of those Kircher had used many years before. He wrote about "Rays of Sound" that reverberate and undulate within the cone, "in the same manner as the Rays of the Sun." To Kircher—who had been experimenting with the use of conical tubes as sound amplifiers since he had installed the eavesdropping tube in his cubiculum—the notion that Morland had invented this speaking trumpet was outrageous. As if his fate rested on being remembered as the inventor of the megaphone, Kircher became determined to show how well his own brand of acoustical tube "might extend itself" if taken outdoors and put to the kind of tests Morland had conducted.

"A tube fifteen palms in length and elaborated with singular zeal" was hauled up to Mentorella, the retreat where on other occasions Kircher sought to reacquaint himself with humility. The "situation of this place was marvelously appropriate and most suitable of all for testing a tube," he wrote, and "at a fitting and peaceful time, both during the day and at night, we tested it."

Underlings notified the people who lived on the "circumambient throng of castles which are discerned from the very peak of this crag and are removed by distances of two, three, four and five Italian miles." When Kircher and others began speaking through the tube "with vehement voice," the people in the hilltop castles of the valley signaled back—in the daytime by raising a curtain or flag, at night by the "combustion of flame"—"that they had distinctly perceived the words one by one."

After this "consummate success," the tube was used to invite all

Speaking trumpets, from Phonurgia Nova

the people within range to services for Pentecost, the feast that comes fifty days after Easter, when the Holy Spirit is said to have descended on the apprehensive disciples. As if "thunderstruck by a voice slipping down from the heavens," Kircher claimed, twenty-two hundred people came. Moreover, "several men, conspicuous in their grandeur, had been stirred by the allurement of the polyphonic instrument from even more remote places; they hastened forth not so much for the sake of devotion as for seeing the tube."

All of this was contained in Kircher's next book, his twenty-seventh, depending on how you counted the various editions. Whichever number it was, the book was the first in Europe devoted exclusively to acoustics, put together almost entirely to make his case against Morland. Indeed, apart from the speaking-tube experiments, much of *Phonurgia Nova* (*New Work on Producing Sound*)—the material on echoes, the properties of sound, and so on—was rehashed from *Universal Music-making*.

"I deem it necessary that the following be made clear to the reader, namely that he not persuade himself that this invention new to this time was brought from England but that around twenty-four years prior it was exhibited in the Roman College," Kircher wrote. "It was this very tube that afterwards was approved, by the stupor of all, for its propagation of voice into spaces most remote with altogether fertile success."

If Kircher wasn't going to get the credit, that didn't mean Morland should. As he'd described in *Universal Music-making*, Alexander the Great was said to have used a giant horn to direct his armies from great distances. There was no telling the degree to which the speaking tube could help in the ongoing fight against the Turks.

THE OLD PRIEST did everything he could to make sure he would be remembered. He began to write the story of his early life. But in 1672, there was more bad news: an assistant to the Jesuit superior general informed him that his museum would have to be moved. Rather than remain in the long, light-filled gallery on the third floor, its home for two decades, it would now have to occupy, as one record has it, "a small dark corridor near the second floor infirmary." The new site was "quite obscure."

Historical accounts suggest the corridor was needed to gain access to the choir of Sant'Ignazio, the Jesuit church abutting the college; it had been under construction for many years. But as Kircher understood it, Jesuit authorities wanted to expand the library. He objected to the move in a letter to the general in May of that year, referring to the original donation of antiquities in 1651. "Thoroughly animated both by a bequest of this kind of such immense and multifarious size and variety and by the established space," he wrote, "it was I who did nothing other than decorate it with worthy magnificence by means of expenses from my own resources, as well as those of my superiors, with pictures and with machines and with other necessary things beyond my poverty.

"It occurred then that with the passage of time . . . through this museum the Roman College acquired a celebrity of reputation so great throughout all of Europe that it seemed no foreigner who had not seen the Museum of the Roman College could say that he had been to Rome."

Kircher pleaded with the general to at least consider modifications to the new space. He made a point of saying that his request had nothing to do with concern for his own reputation.

"For since the place is dark, so that it can have a more rich source of light, the windows must be widened by two palms, and it must be fitted with glass laminas, while also the two bed chambers, which have no use on account of the privation of light, must be broken through and made suitable for housing so many and diverse a multiplicity of objects," he wrote.

"For should these things be done I trust that the museum may maintain its own pristine dignity." But they weren't done, and there's no record of a reply.

KIRCHER'S GRADUAL DECLINE in reputation among Jesuits and among the intellectual elite of Europe (to the extent that they had held him in high esteem) came as others were just becoming acquainted with him.

"Does the conference of learned persons please you?" one travelogue author asked. "See Father Kircher for unknown languages and mathematics."

Kircher passed along observations of Jupiter and Venus to Louis XIV's astronomer royal, and word came back that the French king himself had "deep respect" for his work on hieroglyphics.

Letters continued to arrive from people like Philipp Jakob Sachs von Löwenheim, a doctor in Breslau, and Gaspard de Varadier de Saint-Andiol, the archbishop of Arles. Correspondents wanted to discuss the curative nature of warm mineral springs (were their benefits miraculous or not?), the effectiveness of amulets, and their own investigations into spontaneous generation, carried out, in at least one instance, with fish roe in tubs of milk.

The rector of the Jesuit college of Vilnius wrote to say that he was so beholden to Kircher for so much knowledge and information that he was thinking of adding "Kircherianes" to his name.

Kircher was asked to interpret the hieroglyphic inscriptions on a sarcophagus recently transported from Egypt to Lyon. The result-

ing text, along with Kircher's thoughts on Egyptian burial practices and reincarnation, became the basis for yet another book,
although an uncharacteristically slim one; it was only seventy-two
pages long.

It was for a boy king, Charles II of Spain, that Kircher took up
matters of the Bible. Charles was the nephew and the grandson of
Maria Anna of Spain, who had married the Emperor Ferdinand III;
he was also the grandson and the great-grandson of Margaret of
Austria, a queen of Spain, Ferdinand III's aunt. The product of so
much intermarriage within the Hapsburg family, he suffered not
only from a grotesque example of the Hapsburg chin and lip, but also
from an enlarged head and an oversize tongue that made it difficult
to eat and to speak. Physical infirmities prevented him from walking
until he was eight, and he had some sort of learning disability.
Charles's deficiencies later included the inability to father a child,
and he would come to symbolize the end of the Hapsburg era.
Kircher, now in his seventies, lent an almost childlike quality to his
illustrated volumes on Noah's ark and the Tower of Babel for this
poor young reader.

Even a child might want a book on the ark to address what have
been called "various practical problems" with the story—problems
related to fitting two of every single living species on board a vessel
built by a five-hundred-year-old man and his three sons. To help resolve these, and to try to determine the exact specifications for the
ark that Noah had received from God, Kircher compared the texts
in Hebrew, Chaldean, Arabic, Syriac, Latin, and Greek. Making use
of Galileo's studies of floating bodies to figure the ark's buoyancy, he
created a schematic foldout showing each animal's assigned spot.
This space-planning exercise was supported by detailed speculation

on the care and feeding of the animals, as well as on issues of sanitation. Noah didn't have to include animals that resulted, as Kircher believed, from interspecies mating, such as the mule (the real result of relations between a horse and donkey) and the giraffe (the result of intimacy, he claimed, between a camel and a panther). Deer would have been invited on board, but reindeer didn't need to be, because, as he seemed to suggest, animals have a way of adapting, if not evolving, under different environmental circumstances.

Moreover, there was no need to accommodate lower forms of animals such as mice, frogs, and lizards because those animals arise on their own—they are "born from rot, that is, from the semen of the same animal left somewhere, or from the rotting parts of them." (An exception was made for snakes; they were assigned to the bilge, where they would absorb the putrescent vapors that could lead to disease.) And so the story of Noah's ark, which was not in question, gave credibility to the notion of spontaneous generation. How else to account for all these creatures roaming the earth now? They certainly couldn't have *fit* on the ark.

Kircher was unsure, or unsure for the sake of a boy, about certain legendary animals. No genuine unicorn had ever been located, for example, but to Kircher it was no less improbable than a rhinoceros or a certain horned fish, now known as the narwhal. About the griffin, which was supposed to have the body of a lion and the head of a falcon, he was dubious, though recent reports from China suggested that sometimes eagles or vultures reached a frightening size. Sirens, on the other hand, if not mermaids, were real enough; the siren's "upper part has the sex and appearance of a woman, but its lower portion ends in the tail of a fish," he explained. "There can be no doubt that such a creature exists, for in our museum we have its tail and bones."

20

Immune and Exempt

In 1674, from somewhere in the Palazzo Pitti, and from somewhere under his courtier's wig, Francesco Redi began exchanging letters with an intelligent young mathematics professor at the Collegio Romano, a fellow Florentine named Antonio Baldigiani. It was only natural for Redi to be interested in news of Kircher, whose gullibility about the snake stone he'd sought to expose a few years before. Baldigiani, an admirer of Redi and of the modern approach in general, was happy to oblige.

Kircher "by now is old," he reported to Redi in April of 1675, about a month before Kircher turned seventy-three, "and because of his age, his background, and his history of hard work and in-depth study, he is not always able to be as rational as he would like."

In fact, Baldigiani thought that it would be easy to "play a grand joke on Kircher, who is a perfect target and often comes up against various pranks," though he made a point of not speaking to Kircher about matters of substance. "I am afraid to find myself cited someday in one of his books, as a protagonist or witness to some grand quackery."

Baldigiani told Redi that Kircher had "written a long, rather questionable response" to the public letter Redi had published about the snake stone a few years before. But when he moved to print this counterattack, other scholars at the Collegio Romano intervened, presumably to save Kircher, and the Jesuits as a whole, from the embarrassment of arguing against such overwhelming experimental evidence. Kircher was asked to submit his treatise for a reworking by a venerable colleague: Daniello Bartoli, the author of books on tension and pressure, harmonics, and coagulation, as well as a six-volume history of the Society. But Kircher refused to allow his defense to be revised. "Prof. Kircher is as obstinate as ever in his apprehensions," Baldigiani wrote, "and of the thirty-six books he has printed, from the majority of pages he believes he cannot take out a single line."

Then Kircher became very ill, and for a while it looked as if the matter would resolve itself on account of his death. ("Yesterday morning he had his last communion, and yesterday evening they offered him last rites," Baldigiani wrote later that year. "I, however, do not think him such a desperate case even though his angina is taking him over completely.") But evidently he wasn't ready to die, or, after his recovery, to give up the fight to preserve his legacy. He decided to secure the services of a lawyer, or rather an assistant with legal training, who would be willing to put his own name on a defense against Redi's charges.

This particular disciple, Gioseffo Petrucci, was secular and therefore could operate free of Jesuit authority. He could even publish in vernacular Italian rather than Latin. He'd already participated in a highly visible effort to demonstrate Kircher's greatness: Petrucci was the young student who sent the limited transcription of the Minervan obelisk to Kircher at Mentorella and who was later "thunder-struck" by his interpretation of all four sides.

Since that display several years before, Kircher's reputation as an interpreter of ancient Egyptian wisdom had apparently dwindled within the Collegio Romano to the point where Petrucci was his only student in hieroglyphics. Petrucci was "a solitary and little-known man, dependent in everything on Prof. Kircher," wrote Baldigiani, who was not impressed with him. They had studied philosophy together, and Petrucci had "never shown himself to be even a mediocre scholar." But that may not have mattered much to Kircher, who was essentially going to tell Petrucci what to write anyway.

About two years passed before the product of this collaboration, *Prodomo Apologetico alli Studi Chircheriani* (*Apologetic Forerunner to Kircherian Studies*), saw the light of day. Baldigiani wrote that in Amsterdam even Jansson, Kircher's otherwise enthusiastic publisher, didn't want to print it: "First they pardoned themselves from the work because it was written in Italian and they couldn't read it; then they said that the work was too small and not appropriate for their book-bindings; then they claimed to be too busy with the works of Kircher himself. . . . Finally they agreed to print it after months of correspondence."

In his defense of Kircher, Petrucci backed up the stories that Redi had shrugged off as anecdotal evidence with even more anecdotal evidence: since the experiment with the dog in 1663, he said, Kircher had used the stone to heal another dog, a preacher in Tivoli with an insect bite, and an assistant with an infected arm. On top of this, news of seventeen instances of success with the snake stone had been relayed from the court of Emperor Leopold in Vienna. (The occult-minded Leopold, one of Kircher's greatest patrons, was hardly an objective party. It also happened that Vienna was in the midst of a kind of snake-stone mania that had driven prices for the little stones to absurd levels.) Given all these reports of success, Petrucci wondered whether Redi, rather than Kircher, had been the gullible one: maybe Redi only *believed* he was in possession of genuine snake stones when in fact they were fakes. And wasn't it possible, Petrucci argued, that the hens and other birds that had failed to be healed in Redi's trials were particularly feeble?

As the title of the book suggested, its arguments were meant to apply not just to the snake-stone matter but to "Kircherian studies" in general. In his opening paragraphs, Petrucci described how he'd been stirred to defend his master from the totality of attacks against him—defend him from all the "reckless and impudent slanderers" who had breathed "the pestiferous breath of poisoned invectives" and otherwise made "malignant accusations . . . meant to hinder the perpetual studies of Father Athanasius Kircher, the deserved winner of Glory."

Petrucci conceded that to some extent, after a lifetime confronting the cynicism and the envy of others, Kircher had "by now been made immune and exempt from the stings of zealous Critics, and the bites of malignant detractors, thanks to his great merit, recognized forever, and deserving of inestimable esteem by the most well-known Wise Men of the Universe."

Nevertheless, on the grounds that the truth itself would never consent to the satiric and unfair shaming of his master—to these attempts to "bring down the statue erected to the glory of Father Athanasius Kircher, my most esteemed teacher"—Petrucci intended to clear away the "fumes from the muddy cesspool of ignorance" so that Kircher could live "in the memory of virtuous men," for "all Celebrated future centuries."

As portrayed by Petrucci, Kircher was not a credulous fool but more like a modern skeptic. In the case of the snake stones, even though various priests in India and other parts of the East "constantly insisted on the marvelous virtues of these Stones, and each one of them had their own sensory experience with them," Kircher did not simply believe them. "He did not go according to the sentiments and testimonials he collected, blindly and obediently ceding his will to odd stories," Petrucci wrote. "He did not permit this intellectual fraud through passionate opinions, but kept his mind uncontaminated and unalterable in the quest for truth until a more appropriate moment, in which he would be able to learn from experiment and see for himself."

It was in the face of evidence, *The Apologetic Forerunner* argued, that Kircher distinguished himself from Redi, who Petrucci claimed was too narrow-minded to accept anything but his own preconceived notions about the natural world. "The works of nature are prodigious," Petrucci wrote, "and whoever does not penetrate her reasons imagines these prodigies impossible and does not believe them." For Kircher's willingness to be open to new and surprising discoveries, Petrucci went so far as to compare him to Galileo. Or, since Kircher was behind Petrucci's argument, it was Kircher himself who made the case for the comparison, and who probably believed it. The book quoted many passages from *The Assayer* of 1623, in which Galileo described the experience of coming under constant criticism, an experience Kircher must have recognized as his own. "I have never understood . . ." Galileo wrote, "why it is that every one of the studies

I have published in order to please or to serve other people has aroused in some men a certain perverse urge to detract, steal, or deprecate that modicum of merit which I thought I had earned."

Not many others saw the similarity. Once copies of *The Apologetic Forerunner* finally reached the bookshop of the Collegio Romano in 1677, for example, only two were sold: one to a certain Monsignor Slusio for the archbishop of Prague, to whom the book was dedicated, and one to another patron, at Kircher's request.

21

Mentorella

Despite increasing trouble with his heart and with his hearing, difficulty remembering things, pain from kidney stones, and other frailties and infirmities, Kircher struggled to publish more books before he died. If people wanted experiments, he seemed to decide, then that's what he would give them. A student named Johann Kestler helped cull everything from Kircher's books that could conceivably be called an experiment, and compiled a total of 337 observations and trials into one Latin volume that would be published under Kestler's name: *Experimental Kircherian Physiology in which by the greatest multitude and variety of arguments knowledge of the natural universe is investigated and confirmed by experiments in Physics, Mathematics, Medicine, Chemistry, Music, Magnetics and Mechanics.* It wasn't a coincidence that Kestler's written protestations on behalf of Kircher ("the prodigious miracle of our age") echoed those of Petrucci in substance and in style; Kircher wrote or edited Kestler's remarks as well.

As with all of Kircher's work, the scientific value of each of these individual "experiments," on subjects ranging from the projection of visual images to electrical attraction, varied enormously. Many observations and claims that had been called into question through the years were given elaborate defenses that were themselves unconvincing. But this relatively straightforward selection of Kircher's physical studies—removed from the labyrinthine rhetoric, the utter speculation, and the exhaustive erudition that surrounded them—has caused more than one reader to imagine how much Kircher would have benefited from a good editor all along.

Another assistant worked to prepare a catalog of Kircher's museum to preserve the memory of the collection and the place for which he'd become so famous. An elegant engraving was commissioned for the frontispiece of the book, and it has since become one of the most recognizable images from all of Kircher's printed works. It shows Kircher greeting a pair of visitors within a great, decorated hall. A sunlit array of specimens, skeletons, devices, paintings, and sculptures extends beyond the point where the eye can see. Obelisks, or replicas of obelisks, reach up perhaps four times the height of a man toward cathedral-like ceilings. But the scale represented in the engraving bears little resemblance to the scale of the actual museum in either of its locations within the Collegio Romano. The gallery space the collection occupied until 1672, when it was moved, was nowhere near as vast. And the impressive-looking obelisks were in reality only three or four feet high without their bases. (The obelisks were assumed to have been lost for good before being rediscovered in 1988 in the attic space of the building; the Collegio Romano is now home to a high school.) Maybe what Kircher said about the engravings of ancient sites in *Latium* also applied to this engraving:

it showed the museum "not as it was, but as it could and must have been."

Other projects stalled. Publication of one manuscript, a tour through Tuscany, something like *Latium*, had already been bogged down for many years because of its inaccuracies, and because of sensitivities about how the intellectual history of the region was to be rendered—all the experimentalist studies supported by the Medici, for example, and particularly the contributions of Galileo. At this point the manuscript was passing through various Jesuit hands, being subjected to negotiations behind Kircher's back and revised in places against his will. "These days, because of age," Baldigiani wrote about Kircher, "he has become very difficult to deal with, and also he is very easily bothered and has a tendency to be suspicious."

The volume on Tuscany never saw the light of day, and the manuscript has gone missing. Other books that Kircher had promised or that Jansson had announced on advertising pages bound into his publications—books on such things as Egyptian art and various translations—would never be written. "Many others," a disciple wrote, were "preserved in his mind."

Kircher devoted his last energies to mathematics, a subject that he'd largely overlooked through the years, but that he must have understood was increasingly important to many of the most respected minds of the time as a way of getting at the truth—objective, certain, not debatable. "Philosophy is written in the mighty book that lies forever open before our eyes (I mean the universe), but you cannot understand it unless you first learn the language and the script in which it is written," Galileo had pronounced. "It is written in the language of mathematics." And since then, within a short space of time, people such as Wallis, Napier, Cavalieri, Fermat, Pascal, Mersenne, Huygens, Barrow, Collins, and Roberval had taken the language to a new level.

In devising the coordinate system for plotting curves and other figures on a plane, Descartes brought geometry and algebra together into what is now called analytic geometry. Pretty soon, three-dimensional as well as two-dimensional shapes could be represented by algebraic functions. "A shape in space has given way to an analytic formula," twenty-first-century mathematician David Berlinski has explained. "And with this insight, the first step has been taken in a vast, far-reaching project that will in the end bring all forms of continuous motion, the cannonball *and* rotation of the planets in the night sky, under the control of a numerical apparatus."

Although a professor of "mathematics," Kircher was never a great mathematician. The Englishman John Evelyn had seen him "expound" on "a part of Euclid" in 1644, but when in *The Great Art of Light and Shadow* Kircher claimed to have cracked the age-old mathematical problem of how to square the circle, his naive solution was ridiculed by Mersenne and others. It was perhaps because Kircher was aware of so much activity in mathematics that he thought to publish *Arithmologia*, his 1665 book on the subject of numbers, or rather on the subject of numerology. Somewhat counterintuitively, he seems to have intended that volume to show that he wasn't quite as innocent and foolish as these new mathematicians might believe. He spent a great deal of time laying out the numerological beliefs of "the Cabalists, Arabs, Gnostics and others" in order, he said, to debunk them. And yet for Kircher numbers weren't mere quantities. He believed in the "genuine and licit mystical signification of numbers"—in the Hermetic "Mystic Monad or, if you will, One-ness" that was associated with God and indivisibility, the source and the entirety of all things, and in "the divisions of substance from the divine mind" represented by all other numbers, fractions of the whole.

"There must be no doubt but that within numbers lies hidden a certain proximity to divine nature," he proclaimed. After all, "all creatures breathe numbers: Sky, Earth, Elements and whatever exists in harmony and concinnity with the Angelic, Human, Sidereal and Elemental Universe, all . . . are subjected to the reckonings of numbers."

Now in his late seventies, Kircher began a manuscript composed chiefly of trigonometry computations. If the project represented an attempt to make a real contribution to mathematics, it was a weak one. Just about the time that Isaac Newton and Gottfried Leibniz began quarreling in print over which one of them had been the first to conceive of the calculus, a conception that Berlinski says caused "a reverberating sonic *boom!* in the history of thought," Kircher grew too tired to keep going. He handed the manuscript over to another priest.

IN NOVEMBER OF 1678, Kircher wrote to a colleague: "You must know that now, bowed down by my seventy-seven years of age, I give my time to nothing besides spiritual exercises, nor do I occupy myself with any other studies . . . I am fully occupied in penetrating the science of the Saints, which is to be found in Christ crucified, so that when death comes it will not find me occupied in empty studies." At

The frontispiece of Kircher's Arithmologia

the bottom of the letter there's a postscript: "Please excuse my trembling hand."

Sometime later Baldigiani sent a report to Redi. "Decrepit and old, Professor Kircher is deteriorating at an alarming rate. For more than a year he has been deaf, his vision is failing, he has lost a good part of his memory and he rarely leaves his room, unless to go to the pharmacy or the porter. In short, we have already given him up for lost, though he may live many more years."

Kircher's autobiography, which no one would bother to set into type for more than two centuries after his death, let alone publish in a lavishly illustrated edition, contains some insight into his thoughts in old age. There's a section at the end in which he reflects on the course of his life. It was God, he wrote, who "wished that I expend my small talent . . . for the glory of His divine name and for the benefit of the common weal."

"It was surely God who had destined me from the womb of my mother to pursue this matter in ways marvelous and manifold, first through the efficacy of innate instinct, and then through marvelous and fortuitous happenings, and finally through perils of life endured on land and sea."

And it was surely "He who, although I was ordered to travel to Vienna by the command of my Superiors to serve as professor of Mathematics, led me to Rome in order that indisputably in this theater of the shared universe I devote myself to the explication of the obelisks, a thing finally revealed as my talent, meager as it was."

But it was not possible for Kircher to pretend that he had nothing to do with the trajectory of his life and work, or with the waning of his reputation.

"If only I had accomplished this with such perfection and so great a zeal for the glory of God as my gratitude toward the supreme Father of men was demanding . . ." he wrote. "Yea indeed I sometimes strayed from my deserved end, and drew something to myself from the applause of men that was owed to God alone.

"And so, the subject matter of my studies to which God and obedience have destined me were varied and manifold, and, in the end, hidden and unattainable, indeed an ill-matched task for Herculean shoulders, much less my own, to bear."

IN AUGUST OF 1679, Kircher grew so unwell that he was given the last sacrament again, then regained his strength. Months later a Jesuit professor sent a letter with a question about the telescope. An assistant wrote back with an apology that Kircher couldn't look into it himself; an astronomer at the Collegio Romano would reply.

By March of 1680, Kircher had begun what a fellow priest called his "second childhood," which lasted months. His condition began to worsen in November, about the time the seventy-nine-year-old Bernini suffered what must have been a stroke: his whole right side became paralyzed and he lost the ability to speak.

Kircher died on November 27. Bernini died the very next day. The result was that funeral services for Kircher were overshadowed by the tremendous outpouring of tribute to the baroque master.

KIRCHER'S BODY WAS to be entombed in the great Jesuit church of Il Gesù, just a few narrow streets away from the Collegio Romano, but he left instructions for his heart to be taken to Mentorella and buried at the foot of the altar there.

Preparation for burial in the seventeenth century began by cutting open the chest for the removal of internal organs. The cranium was also sawed open, and once it was embalmed, Kircher's brain, the source of all of his ideas and all his trouble, was packed in a barrel with his intestines, eyes, tongue, lungs, liver, and other organs. His heart was set aside. Deep incisions were made all along the limbs, the back, and the buttocks to drain the blood from the veins and arteries. After all the viscera, layers of fat, and tendrils of membrane were removed, the open cavities were scrubbed with spirits of wine, turpentine, and aromatic oils. Then they were filled with untwisted cotton or tow and large quantities of embalming powder, which was like finely ground potpourri. One recipe alone included rosemary, laurel, hyssop, absynth, mint, rue, sage, wild thyme, pennyroyal, oregano, germander, lavender, chamomile, fennel, rosewood, spikenard, caraway, angelica, cloves, valerian, aloe, mastic, incense, myrrh, styrax, labdanum, canella, mace, and saffron.

The body was sewn up, bathed in spirits and liquid balm, and anointed. Before being placed in a casket, it was bandaged, dressed in vestments, and wrapped in a linen sheet, which was tied with ribbon at the head and feet. When the time came, Kircher's wooden coffin was plastered into a lengthwise space in the rough wall of the crypt beneath Il Gesù, the place where aboveground the body of Ignatius of Loyola and the right arm of Francis Xavier are opulently interred.

Kircher's heart was removed from its fibrous sac, soaked in spirits, and thoroughly cleaned. The ventricles were filled with embalming powder. Then the heart was anointed with oil, essence of nutmeg, or tincture of musk. It was set in perfumed cotton, powdered, and placed inside a little waxed bag. The bag was put in a small box. The box was wrapped in violet taffeta.

Perhaps only a few Jesuit companions traveled with this box to Mentorella. The trip from Rome to Tivoli to the mountain took several hours by cart or carriage. From there, the way to the shrine—up the mountain, through the little village of Guadagnolo, and then along the summit to the cliff-side spot—was an ordeal in itself. People now can ride to the top in a car, but for hundreds of years the most common mode of transport there was mule.

"The track which leads to it from the valley, at the base of the mountain, can hardly be called a road, though there is no other direct way of reaching it," explained an English priest who made the pilgrimage in the late nineteenth century. "A mule is the only animal that can be trusted to carry one there in safety. It is amazing to see with what steadiness that sure-footed animal trips along the ledge of a precipice, where one false step would be fatal to it and the rider. But the danger of the ascent is soon forgotten when one beholds the beauty of the approach to the sanctuary and the grandeur of the scenery for many miles around."

It's fair to say this last climb had little effect on Kircher's dead and well-protected heart, but on previous ascents it must have been stirred. The trip may have put him in mind of his herniated journey over the mountains of the Hochsauerland to Paderborn, or any number of climbs in Malta or Sicily.

The Jesuits who brought his heart up to the mountain said Mass in the little basilica that Kircher had restored, gathering in view of the wooden Madonna who had seemed to address him on the day he discovered the church. They set the box in a space beneath the marble at the foot of the altar, and then sealed it up before returning to Rome, where life went on as usual.

EFFORTS WERE MADE over the years to restore Kircher's museum at the Collegio Romano to its former glory, but in the meantime no one kept very good watch over the items and machines that he had painstakingly collected. As a 1709 account has it, there was a clerk "kept quite busy with other undertakings and thus not fully capable of caring for them as well as he should have." Some items were broken or worn by misuse, and some were not maintained. Others "simply vanished after exposure to the eyes of visitors apart from the guard."

A French traveler, visiting Rome several years after Kircher's death, reported that "Father Kircher's Cabinet in the Roman College was formerly one of the most curious in Europe, but it has been very much mangl'd and dismember'd."

22

Closest of All to the Truth

I t's hard to say exactly when the modern age began. But if "what the modern world's about, what it *is*, is science," as David Foster Wallace wrote in his history of the concept of infinity, and modern science is "essentially a mathematical enterprise," then the modern age has something to do with Isaac Newton. Newton's *Philosophiae Naturalis Principia Mathematica*, published in 1687, seven years after Kircher's death, offered an astoundingly rigorous mathematical proof of the Copernican system. Along the way it introduced the notion of absolute space and time (although that was later undone by Einstein's theories of special and general relativity), laid down the fundamental laws of motion (based on the work of Galileo, Kepler, and Descartes), and explained the movement of the planets, as well as of the comets, the moon, and the tides, by virtue of a single, universal, gravitational force.

The title itself, translated as *The Mathematical Principles of Natural Philosophy*, seems intended to signal the end of the confusion that existed throughout Kircher's lifetime between what it meant to demonstrate something (or prove something wrong) and what it meant merely to speculate. "I hope to show—as it were, by my example—how valuable mathematics is in natural philosophy," Newton later stated. "Instead of the conjectures and probabilities that are being blazoned about everywhere, we shall finally achieve a natural science supported by the greatest evidence." Privately Newton was tired of the notion that mathematicians "who find out, settle & do all the business must content themselves with being nothing but dry calculators & drudges," while others do "nothing but pretend and grasp at things." Rather than providing superficial quantities and measures that shed no light on the deeper truth and essence of nature, mathematics in Newton's hands seemed to get *right* at its essence, leading to the popular Enlightenment notion that God himself was a sort of mathematician who put an elaborate clockwork universe into place.

The *Principia* was printed with the permission of the Royal Society but paid for out of the pocket of Edmund Halley, later of comet fame, because the society's only other publishing venture, the lavish *History of Fishes*, had been a flop. Some time naturally had to pass before the arguments of the *Principia* were widely accepted. Very few people could even understand them. This was despite the fact that Newton had tried to make them as accessible and palatable as he could. He had originally reached his conclusions by means of his own "infinitesimal calculus," but for publication he'd gone back and made his arguments work within the more widely understood laws and language of geometry.

Newton's subsequent reputation was such that a hundred fifty years later—in the midst of the industrial age, which owed its existence to Newtonian mechanics—an introduction to a new edition of the *Principia* could state that it "is celebrated, not only in the history of one career and one mind, but in the history of all achievement and human reason itself." The very character of its author was "held up as the noblest illustration of what Man may be, and may do, in the possession and manifestation of pre-eminent intellectual and moral worth."

In day-to-day life this "noblest illustration of what Man may be" was variously awkward, ambitious, obsessive, easily offended, vindictive, petty, and mean. A troubled celibate who was at least *said* to have laughed only once or twice, Newton used his wealth later in life to upholster in crimson much of what he owned. He was born the year Galileo died, and entered Trinity College, Cambridge, about a year before Kircher turned sixty. Quickly dismissing the Aristotelian doctrine still being taught, he read ancient and modern mathematics based on his own curiosity, and was hugely influenced by the new method and physical philosophy of Descartes as well as Gassendi. In 1665, when the plague came through England, Cambridge was closed down, and he stayed at his family home in Woolsthorpe, Lincolnshire, with nothing but time on his hands.

During a two-year period of manic productivity, Newton (literally) figured out what today is called the generalized binomial theorem and developed his basic method of fluxions and fluents, his terms for differential and integral calculus. (As he described it, this was the method by which, "given the length of the space continuously" at every instant, he could "find the speed of motion at any time proposed," and, inversely, how, "given the speed of motion continuously," he could "find the length of the space described at any time proposed.") His optical investigations included sticking a bodkin or blunt needle underneath his eyeball to see how changing its curvature would affect what he saw. Through dozens of experiments with prisms of his own manufacture, he determined that colors weren't produced by alterations to light, as everyone assumed, but that white light was *made* of colors. And whether or not a falling apple in the Woolsthorpe orchard was as crucial to Newton's thinking as the famous story suggests, he "began to think of gravity extending to ye orb of the Moon," as he later claimed, and calculated (from Kepler's laws on planetary motion) that the forces required to keep the moon and the planets in their orbits would operate in "pretty nearly" the same way as the force that pulls an apple to the surface of the earth. That is, according to the inverse of the squares of their distances from the larger bodies.

It's possible to argue that Newton got more right between the ages of twenty-three and twenty-five than Kircher got right in a lifetime. But their interests, outlooks, and erudition, as well as many of their ideas, if not their methodologies or their mathematical abilities, were strikingly similar. Early in his college career, Newton made sundials and clocks, experimented with magnets, tried his hand at creating a universal language, and "ground and polished glasses . . . for all kinds of optical purposes." When he studied color, his reading of ancient harmonics (the "harmony and discord of sounds") led him to look for a correspondence between the colors of the spectrum and the seven notes of the diatonic musical scale. (More accurately, he looked for a correspondence between the *ratios between the notes* and the *ratios between the colors*.) Cheating a little with his math to find one, Newton added a seventh color to the rainbow that doesn't really belong: indigo, between blue and purple.

Years later, Voltaire heard that Newton had gotten his ideas about the correspondence between color and sound from *The Great Art of Light and Shadow*. But after he cracked open the book to a section where Kircher matched types of human voices to colors, and to animal natures, and to personality traits, he dismissed the idea as laughable. It's true that Kircher's name is absent from Newton's long, handwritten lists of authors and authorities, and from the catalogs of books in his library. Given the fact that Kircher was read by virtually every other intellectual in England at the time, however, and that Newton's library contained della Porta's *Natural Magic*, Agrippa's *Occult Philosophy*, the body of the works said to be written by Hermes Trismegistus, and more than a hundred fifty other books that might be described as magical or alchemical, its absence is conspicuous. There is no way to know if Newton read Kircher, but it's very likely that he did.

Once Newton returned to Cambridge, he built the first reflecting telescope, more powerful despite its tabletop size than the 150-foot refracting telescope of the Polish brewing heir Johannes Hevelius. Since it operated by virtue of mirrors, it might have been called, as Kircher might have called it, the first catoptric telescope. Around the same time, he also set up a laboratory in a shed adjacent to his room, with glassware and furnaces for chemical and alchemical experiments. He subsequently pursued alchemy with incredible, meticulous intensity for more than thirty years, not quite to the exclusion of everything else, secretly becoming, in a modern biographer's words, "the peerless alchemist of Europe." Newton's assistant recalled: "The Fire scarcely going out either Night or Day, he siting up one Night, as I did another, till he had finished his Chymical experiments. . . . What his Aim might be, I was not able to penetrate into, but his Pains . . . made me think at something beyond the Reach of humane Art and Industry."

Newton steeped himself in the notion of a pristine original faith.

In pursuit of sacred knowledge, he read a vast range of texts by ancient and esoteric authors, and spent decades trying to piece together a precise chronology of humankind and its theologies. (More than half of Newton's written output is on theology and religion, and he wrote or copied down more than a million words on alchemy.) Newton studied the texts attributed to Hermes Trismegistus with increasing attentiveness and respect over his lifetime, regarding the alchemical tracts attributed to him as older and closer than any others to original wisdom. And he struggled to reconcile the new math-based mechanical philosophy with Hermetic descriptions of a primordial power that, as worded in his own translations, "penetrates every solid thing."

Something appeared to be at work in many natural processes—an "active principle," a force—that could not easily be explained by the movement of perfectly dead particles of physical matter. Newton's alchemical experiments, his tests on air and electricity, and a great deal of his thinking, were aimed at getting to the bottom of it. Many chemical processes could be explained by what he called "vulgar" or "brute" physical interaction, the "gross mechanical transposition of parts." But how to explain, for example, the spontaneous generation of living vermin from rotting matter without some animate presence, living agent, or spirit? (Newton never doubted that living vermin were engendered this way. Francesco Redi wrote *Experiments on the Generation of Insects* in Italian, not Latin, and besides, the book didn't settle the debate.) It was increasingly unsatisfactory to say, as the Aristotelians did, that it was the nature or final cause of an acorn to become an oak tree. But then what *did* cause the acorn to become an oak? If Newton's preliminary calculations about the orbit of the moon were correct, how was it that the Earth could pull at the moon from such a distance with no physical contact whatsoever?

In an early alchemical paper, Newton described the process of generation and putrefaction by which he believed metals, and other living things, *vegetated*: it was the "sole effect of a latent spirit" that had penetrated or become intimately embedded within "rude matter" in various states of "maturity." This "vegetable spirit," which he also referred to as "Nature's universal agent, her secret fire" and "the material soule of all matter," was unimaginably subtle, traveling around in vapors and fumes through such processes as fermentation, putrefaction, and evaporation, and participating in various material forms: "When two vegetable spirits are mixed of unequall maturity," Newton wrote, "they fall to work, putrefy, mix radically & so proceed in perpetuall working till they arrive at the state of the less digested." In his alchemical shed, Newton was keen to record instances of attraction and repulsion at the smallest observable level. About certain chemical compounds he noted a "propensity to associate with one another," about others "a secret principle of unsociableness."

About two years after Kircher published *The Magnetic Kingdom of Nature*—in which he emphatically reiterated his notion that "the world was bound with secret knots," and insisted that the "hidden form operating in all things" should be called "magnetism"—Newton began referring to the active force as "magnesia." "This & only this," Newton wrote, "is the vital agent diffused through all the things that exist in the world."

MEANWHILE, AMONG MEMBERS of the Royal Society such as Christopher Wren, John Wallis, John Wilkins, and Robert Hooke, consensus was increasing around the idea first proposed by William Gilbert in *De Magnete*: that a magnetic or similar kind of connection was responsible for keeping the moon and the planets in motion, and from flying out into oblivion. "Whether by any Magnetick or what-other Tye," Wallis stated, that connection was "past doubt." Hooke wrote a paper as early as 1666 "concerning the inflection of a direct motion into a curve by a supervening attractive principle," and later engaged Newton in a series of exchanges on the idea. He shared his own theory—that a drawing power of this sort would be inversely proportional to the square of the distance between two given bodies. But it was only Newton who could do the math, demonstrating, for example, that such an attraction would produce Kepler's elliptical orbits.

Newton didn't have much trouble determining that this universal force wasn't actually magnetic, since "some bodies are attracted more by the magnet; others less; most bodies not at all," and since the sun "is a vehemently hot body & magnetick bodies when made red hot lose their virtue." But the gravitational force described in the *Principia* and associated with Newton's greatest scientific contribution was nevertheless an invisible attractive power—the kind of unseen and, as far as Newton could determine, immaterial force that Kircher would have said he'd *already* described, long before.

Gravity was "occult," at least in the sense of the word's earliest meaning. But it was also, as Newton put it, "manifest." In the end, he could quantify and measure it by its effects, but he could not say exactly how or why it worked. "I have not been able to discover the cause of these properties of gravity from phenomena, and I feign no hypotheses: for whatever is not deduced from the phenomena must be called a hypothesis, and hypotheses, whether metaphysical or physical, or based on occult qualities, or mechanical, have no place in experimental philosophy," Newton tried to explain in the second edition of the *Principia*. "It is enough that gravity really does exist and acts according to the laws that we have set forth, and abundantly serves to account for all the motions of the celestial bodies and of our sea."

But in many respects, it wasn't enough, least of all for Newton, who almost made himself crazy wrestling with questions about the cause of gravity. (In truth, his apparent breakdown in 1693 was more likely the effect of exposure to so much mercury in his alchemical shed.) He obsessed over the possibility that a very subtle material or semi-material explanation might be found after all, and over his inability to take his discoveries further. "For many things lead me to suspect," he'd written, "that all phenomena may depend on certain forces by which particles of bodies, by causes not yet known, either are impelled toward one another and cohere in regular figures, or are repelled from one another and recede." At least Newton was content in his belief that the ancients had understood everything he'd found out, and much more besides. He believed he'd been able to restore some part of the old knowledge that had been lost along the way.

Newton's statement wasn't remotely enough for a number of the most fiercely intelligent intellectuals of the time. For people like Christiaan Huygens and Gottfried Leibniz, gravity had less to do with new mathematics than with, for example, the old salve that could heal the injured when applied to the weapon that had wounded them. Leibniz and Newton were still in the midst of their ugly dispute over who had been the first to develop calculus; they went at it for many years, publishing vicious attacks on each other under false

names. Though an advocate of unity and accord in almost every political, philosophical, and religious circumstance, Leibniz felt intense
enmity for Newton. To him, Newton's gravity was an "occult quality" in the modern sense: it was magic. It was, he said, "a supernatural thing, that Bodies should attract one another at a distance,
without any intermediate means."

Leibniz was not actually against the idea of immateriality. He
later came to understand the world as utterly immaterial, mere perception, a phenomenon composed of aggregations of pure soul-like
substances. Borrowing from the Pythagoreans, he called them
monads. These were "the real atoms of nature," he said, which formed
divine, sophisticated, lifelike agreements with one another through
"pre-established harmony." In Leibniz's conception, there was no
need to distinguish between body and soul because even though the
"material" world was real enough to experience with the senses and
to be "explained mathematically and mechanically," there was no actual body.

Nevertheless, Leibniz's argument against Newton's conception of
gravity was that it was an "immaterial and inexplicable virtue." Not
only was it "invisible, intangible," and "not Mechanical," it was "unintelligible, precarious, groundless, and unexampled." From Leibniz's
point of view, it "must be a perpetual *Miracle*: And if it is not miraculous, it is false."

Whatever gravity actually is—Einstein said it was not really a
force, but a warp in the dimension of space-time; in the twenty-first
century, scientists are somewhat hard-pressed to say—it came to be
recognized as fact. And, of course, Kircher came to be associated
with the fictions of the pre-scientific past. "Surely it is no coinci-

dence," a modern historian says, "that the crystallization of Kircher's reputation as the most ridiculous of the late Renaissance encyclopedists and the emergence of Newton as the first man of science both occurred in the same period." Pretty soon their fates were sealed.

But even a century after Kircher's death, there were still a few people for whom the distinction wasn't so clear-cut. "In my opinion the Egyptian system of the world, which was based on the laws of attraction and repulsion, seems to be the closest of all to the truth," wrote a Slovak intellectual named Adám Ferencz Kollár in 1790. "This opinion of mine now has the consent of all Europe, which approved it not so long ago, but attributed it to Newton, in his calculus. But Kircher came before Newton; and lest someone thinks that I am daydreaming, I would have him read carefully and with an unprejudiced mind those things that Kircher wrote."

23

The Strangest Development

A lthough Leibniz, for his part, had Kircher to thank for much of his early thinking and inspiration, over time he changed his opinion of Kircher's abilities, and in the end took apparent pleasure in scoffing at him. He'd sent Kircher oily praise in 1670 for *The Great Art of Knowing*, but by 1716 he'd decided that Kircher "had not even dreamed of the true analysis of human thoughts." The same year, about Kircher's interpretations of the hieroglyphs, Leibniz offered a final, offhanded conclusion: "He understands nothing."

But it was largely because of Kircher that Leibniz became interested in Chinese culture, philosophy, religion, and language. (Kircher's *China Illustrated* was published the year Leibniz turned twenty-one.) And if it hadn't been for Leibniz's interest in China, and his own subsequent correspondence with a number of Jesuit missionaries, he never would have found what he believed to be the ancient precedent for his system of binary arithmetic. In 1701 a priest in Peking sent Leibniz a treatise on the *I Ching*, the text now popularly used for divination and the interpretation of events. It was composed of sixty-four hexagrams and said to be authored by an ancient Chinese philosopher-king. When Leibniz examined the sequence of hexagrams, he decided almost immediately that it was a rendering of binary numerical progression. (It wasn't.) In his view the *I Ching* was perhaps the very first mathematical-metaphysical text. And because he'd been persuaded by Kircher's arguments that Chinese culture had descended directly from that of the ancient Egyptians, he believed he knew the true identity of the king who had written it down: Hermes Trismegistus. With this assurance that he'd restored some part of the original wisdom, Leibniz published his description of the binary scheme in 1703, and made it so that, at least according to his conception of it, every bit (short for "binary digit") of information in every modern digital device contains some combination of God (1) and nothingness (0).

Kircher's influence often worked this way: in spite of the negative opinions of his readers, and sometimes in spite of himself. His incorrect assertions about the Egyptian roots of Chinese society were responsible for the fact that, for more than a century after his death, many serious scholars tried to unlock the secret of the Egyptian hieroglyphic system . . . by studying Chinese. But when Jean-François Champollion made his breakthrough with hieroglyphics (after the discovery of the Rosetta stone by Napoleon's forces in Egypt), he did it with Kircher's help.

More precisely, he did it with the help of Kircher's Coptic grammar and lexicon, something that Kircher himself had largely neglected in his attempt to read messages he assumed somehow transcended mere linguistics and were mystical in form as well as meaning. As it turned out, Kircher's hapless French mentor, Nicolas-Claude Fabri de Peiresc, had been right all along to push for an understanding of the language of the early Egyptian Christians. But without the Rosetta stone—which was inscribed with the same decree in three scripts: hieroglyphics, a later form of Egyptian script related to Coptic, and Greek—even Champollion might not have uncovered the somewhat mundane, phonetic component of the problem. Once the hieroglyphics were at last accurately deciphered, one historian writes, "the obelisks were seen to enshrine not 'the highest mysteries of Divinity,' as Kircher thought, but rather a dull record, for the most part, of the acts and attributes of kings."

This is not to say Kircher's work on Egypt went entirely by the wayside, or was completely forgotten. As with much of what he did, it just wasn't remembered the way he'd hoped. Kircher is still sometimes called the father of Egyptology, though as such, as with his Chinese studies, he played at least some role in creating Eurocentric perceptions about the East—had something to do with the creation of the exotic Other that, as Edward Said argued in *Orientalism* in 1978, went hand in hand with cultural and political power, imperialism, and colonialism.

Otherwise, the great dubious contribution of *Egyptian Oedipus* was that it served as a reference work on the so-called sacred sciences and occult practices. Kircher included many halfhearted disclaimers for his own sake and for the sake of the censors, but his long considerations of just about every magical and mystical tradition helped preserve them for future study. In some cases he had an effect on the traditions themselves.

In some respects, Cartesian dualism, the boundary drawn between body and spirit, helped make the world safe for religious and spiritual practices of all kinds. If occult virtues were no longer legitimate explanations for natural phenomena, then physical science was often seen not to apply, or rather to fall short of applying, to mystical matters. Kircher's books, which provided at least some fodder for the development of the secret societies of the seventeenth and eighteenth centuries, played an even greater role in the nineteenth century, when a fascination with spiritualism and parlor pastimes like spirit-conjuring and levitation emerged. His tree of life and many of his concepts of Kabbalah were adopted almost wholesale by such societies as the Hermetic Order of the Golden Dawn, in which the poet William Butler Yeats and his wife, George, were later active. (Yeats also tried to re-create the vegetable-phoenix phenomenon that Kircher had supposedly demonstrated for Queen Christina, but was unsuccessful in producing an apparition of a flower from its ashes.)

The self-proclaimed psychic known as Madame Blavatsky built her career arguing, in effect, against the Cartesian split. In her view, *both* modern science and Christianity existed in arrogant isolation from larger occult truths—truths she said were well understood by (who else?) the ancients. This mysterious Russian woman became a celebrity of sorts after she arrived in New York in the 1870s. At the salons and séances held in her apartment, she espoused an all-encompassing approach she called Theosophy (which means "divine wisdom"). In what might be called the Kircherian tradition, Blavatsky published impossibly erudite tomes such as *The Secret Doctrine*, in which she revealed the knowledge of the ancients and certain Eastern cultures, attempting to synthesize or at least to analyze every strain of spiritual and scientific belief held throughout human history. As it turned out, her erudition *was* more or less impossible; to write her books, she cribbed from encyclopedic nineteenth-century histories of these teachings, some of which had been taken, at least in part, from Kircher. She even quoted Rabbi Barachias

Nephi, the author of the manuscript that Kircher may or may not have invented, and which he never got around to translating. Blavatsky is generally credited, if that's the right word, with providing the foundation for the New Age movement of the twentieth century.

To her, Kircher was a "monk" who "appeared among the mystics" with a complete philosophy of universal magnetism. "He asserted that although every particle of matter, and even the intangible invisible 'powers' were magnetic, they did not themselves constitute a magnet. *There is but one* MAGNET *in the universe, and from it proceeds the magnetization of everything existing.* This magnet is of course what the kabalists term the central Spiritual Sun, or God. The sun, moon, planets, and stars he affirmed are highly magnetic; but they have become so by induction from living in the universal magnetic fluid—the Spiritual light."

KIRCHER'S MAGNETIC PHILOSOPHY must have been a source of inspiration to the physician named Franz Anton Mesmer, whose elaborate, hypnotic cures became the rage among fashionable Parisians in the late 1770s. Mesmer studied at Jesuit universities in Bavaria and learned the art of magnetic medicine from a member of the Society in Vienna named Father Hehl (sometimes rendered in English as Father Hell), an education that must have drawn on Kircher's works as well as the old magnetic literature. Mesmer came to believe that mere magnets, vehicles of "mineral magnetism," as he put it, were insufficient to treat certain ailments, especially "nervous" afflictions now known as mental or psychological illnesses. He thought many sicknesses were caused by blockages and imbalances of a much more subtle, universal magnetic "fluid" that ran through all things, living and nonliving, by virtue of a force he called "animal magnetism." He used Newton's theories of universal gravity to bolster his arguments. Just as the moon's gravitational pull on the oceans caused the tides, he said, the movement of planets caused changes in the levels of this invisible fluid within the body. (Animal magnetism, he wrote, acts "at a distance" on a principle of "Flux and Reflux," though he might as well have described it as attraction and repulsion, sympathy and antipathy, or consonance and dissonance.) Mesmer developed new kinds of treatments based on his belief that this force could be "communicated, propagated, stored, accumulated, and transported," by sound, light, touch, and even thought.

After meeting with skepticism and some high-profile failures in Vienna, he moved to Paris. The French—young French women, in particular—were somehow much more responsive to his therapies. Soon his house was so crowded with patients that he began to administer to groups of them at a time in multisensory healing rituals. There was incense, Aeolian harp music, mirrors, and colored light. Patients gathered around Mesmer's strange apparatus—a great tub filled with "magnetized" water, out of which protruded movable iron rods that were applied to afflicted parts of the body. Wearing a long lilac robe and "a look of dignity," he lingered with each patient, staring deeply "to magnetize them by the eye." He stroked them "with his hands upon the eyebrows and down the spine; traced figures upon their breast and abdomen with his long white wand." Mesmer's protégés meanwhile gave individual care to the others. They embraced them and rubbed them "gently down the spine and the course of the nerves, using gentle pressure upon the breasts of the ladies."

His patients, thus "mesmerized," as it came to be known, responded by going into reveries or into convulsions, sometimes by sobbing or screaming or laughing uncontrollably. Many said the treatment made them feel better. (Critics said that while Mesmer was successful with the young ladies, he couldn't seem to cure his own long-suffering wife.) Mesmer insisted that his therapy worked by acting on the magnetic fluid within his patients. He may actually have been hypnotizing them. In fact, although the concept of animal magnetism was eventually discredited, his techniques led to the development of therapeutic hypnosis. Disciples and others who practiced offshoot forms of his treatments learned to build what Mesmer called *rapport*, or harmony, with their patients, and began to concentrate on the role of the mind and the emotions in certain illnesses and their cures. As a result, almost every history of modern psychotherapy begins with a study of Mesmer.

Beyond this, although the influence of Kircher's magnetic studies and philosophy has faded, magnetism itself has taken on ever-increasing significance in almost every scientific and technological field, especially after a relationship between magnetism and electricity was discovered in the 1800s. "Without the stunning progress made during the last several centuries in understanding the nature of magnetism, our modern technological civilization would not yet have come into existence," an American professor of astrophysics named Gerrit Verschuur wrote in 1993. "Every facet of the civilized world rests, ultimately, on the widespread availability of electricity to drive the machines of industry. We would never have learned to produce electricity if it were not for the profound insights that arose from the study of magnetism." It isn't just that magnetism is employed in the everyday generation of electricity, or that, for instance, satellite transmissions, cell-phone calls, and wireless connections exist as a flow of electromagnetic waves—that massive amounts of data are routinely transported from one given physical location to another, electromagnetically, invisibly, through the air. Light itself is electromagnetic in nature, as is every interaction between atoms, and almost every physical phenomenon besides gravity.

The realm of unseen energies that Kircher and many of his contemporaries imagined may not exist, but there is nevertheless a realm of unseen energies. In the late twentieth century, astronomers and cosmologists came to a bizarre conclusion: Only four percent of the universe is made up of stuff we understand. Twenty-three percent is made up of something known only by its gravitational effects that they call "dark matter." Seventy-three percent is made up of some kind of antigravitational force they call "dark energy"; it is similar to dark matter, but, in the words of one cosmological theorist, it is "more energy-like." Maybe it wasn't completely and totally wrong for Kircher to suggest that the world is bound with secret knots.

Usually, of course, the practice of science makes the world seem less, not more, mysterious. Today, for example, evolutionary biologists are pretty well convinced that human consciousness is a Darwinian adaptation—that awareness and sense of self evolved over a very long period of time like everything else. In which case perhaps so did the feeling that some part of us might live after the body dies. "Proof will require a lot more information about, for example, neuro-circuitry and the nature of memory and emotional inputs in reasoning," eminent biologist Edward O. Wilson explained in 2002. Nevertheless, he thinks "the Cartesian notion of dualism between body and soul is dead forever. I'm sorry, but that's the way it is."

Life itself may always be a mystery, if not a miracle. As it turns out, Francesco Redi's controlled experiments with decaying meat didn't end the debate about spontaneous generation, or the experimentation, which was conducted on increasingly microscopic, microbial, and bacterial levels for a few more centuries by scientists such as Louis Pasteur. In the second part of the nineteenth century, arguments for spontaneous generation became an integral part of the debate over evolution. It was the Darwinists—the modern scientists, the materialists—who found they had to contend with the problem of how life could be engendered from non-living matter. (Maybe every modern Darwinist still does.) One science historian expressed it this way: "To believe in evolution and a completely naturalistic world-view required the belief that . . . living organisms must have been capable of arising from nonlife at least once on the early earth."

PEIRESC PUT IT very mildly when he said that Kircher's ambitions were "a little grander than the ordinary goals of his colleagues." This led to a lack of restraint as well as other problems, including a certain flexibility with the truth. But for Kircher there were greater truths and lesser ones; there were different measures of truth, metaphors, and multiple meanings, things for which fact-based modern science has no place. Progress required another kind of split, between the literal and the literary. But that was not a split Kircher ever would have been able to abide. And it makes sense that as his scientific reputation diminished, his work continued to capture and to fuel the creative imagination.

The baroque poetry of the Mexican nun Sor Juana Inés de la Cruz was inspired by the Hermetic language in Kircher's books. As a librarian at the Bibliothèque Sainte-Geneviève between 1913 and 1915, Marcel Duchamp examined the optical devices in Kircher's *Great*

Art of Light and Shadow. Giorgio de Chirico's illustrations for Jean Cocteau's *Mythologie* owe a debt to the engravings of the Great Flood in Kircher's *Noah's Ark.* Even the Eye of Providence on the reverse of the Great Seal of the United States, the one above the pyramid on the dollar bill, appears to have been drawn from the frontispieces of books such as *The Magnet, The Great Art of Knowing, Universal Music-making,* and *The Tower of Babel.* Whether Kircher first found it in a book by Robert Fludd or in an ancient or apparently ancient or another source, millions of people now carry this all-seeing eye around with them in their pockets.

In Edgar Allan Poe's story "A Descent into the Maelström," the narrator comes face-to-face with a mile-wide vortex in a northern sea, and is understandably awestruck. "Kircher and others imagine that in the centre of the channel of the Maelström is an abyss penetrating the globe, and issuing in some very remote part," Poe says. "This opinion . . . was the one to which, as I gazed, my imagination most readily assented."

Jules Verne's *Voyage au Centre de la Terre* (*A Journey to the Center of the Earth*), first published in 1864, had many influences. But the German professor, a "learned egoist," who takes the famous subterranean trip, along with his nephew and a guide, bears a striking resemblance to the author of *Underground World.* The story begins when this linguist, mineralogist, mathematician, and museum curator deciphers a coded message found within a runic manuscript. Actually, his nephew deciphers it; as in the story of a trick played on Kircher, it turns out merely to be Latin written backward. The message reveals a volcanic crater on Iceland as the entryway to the realm below—for which they leave the very next day, and into which they descend, hiking through passageways, traveling by raft on underground rivers and a hot ocean, meeting with adventure and discussing the geological theories of the day, for more than two months, until they are lifted on a gigantic wave and forced out through a venthole onto the sunny volcanic slopes of Stromboli, in view of Sicily.

Roberto della Griva, the main character of Umberto Eco's novel *The Island of the Day Before,* isn't sure what to make of Father Caspar Wanderdrossel, the German Jesuit professor he meets aboard the ship on which he finds himself marooned. Was he "a sage? That, certainly, or at least a scholar, a man curious about both natural and divine science. An eccentric? To be sure." Roberto had learned, in Paris and Provence, from Pierre Gassendi among others, to be skeptical about the kind of miraculous stories Wanderdrossel told. But he'd also learned "to concede only half of his spirit to the things he believed (or believed he believed), keeping the other half open in case the contrary was true." Almost everything the Father said was "most uncommon," Roberto admits. "But why consider it false?"

The truth does have a way of shifting over time. Kircher wrote in the preface of *Ecstatic Journey* that there has scarcely been an age of human beings that hasn't "gladdened the World to the extreme . . . with the spectacle of its own new divine power." After all, "venerable antiquity never knew anything about the existence of the new World; it knew nothing about the diffusion of Oceans around the Orb of lands; it had discovered nothing about . . . a great variety of exotic things." And "if anyone had told these things to the ancients . . . they would hesitate even to imagine them." The achievements of each generation have led it into "love and admiration for itself." This phenomenon had "occurred most powerfully" in own his time, "with the great amazement of mortals," and he understood it would continue to occur, again and again.

In the same way that many of Kircher's misconceptions are really misconceptions only from a modern point of view, at least some of our own greatest certainties will be seen as laughingly obvious errors by people, if people are still around, three or four hundred years from now.

In the meantime, it's clear that the modern perspective is simply not the right one to take when it comes to Kircher and to the entire, incredible Kircherian enterprise. There's something to be said for his effort to know everything and to share everything he knew, for asking a thousand questions about the world around him, and for getting so many others to ask questions about his answers; for stimulating, as well as confounding and inadvertently amusing, so many minds; for having been a source of so many ideas—right, wrong, half right, half-baked, ridiculous, beautiful, and all-encompassing.

GIOVANNI BATTISTA RICCIOLI, the Jesuit astronomer in Bologna to whom Kircher forwarded celestial readings for many years, thought Kircher deserved to be remembered. In 1651, within a book-length argument against the Copernican system (a book that was itself otherwise bound to be forgotten), Riccioli published an important set of maps of the moon. They introduced what eventually became the accepted scheme of names for major features of lunar topography. The surface of the moon was not even remotely pristine; it had mountains and valleys and "seas," as he decided to call them, and many, many pockmarks. Riccioli gave names to almost two hundred fifty craters and other sites. He named them for saints, ancient philosophers, Greek gods, and Roman emperors, but also for major astronomers, intellectual figures, and Jesuit scholars of his own time. Kircher was on the list.

The crater named after him can be seen in an area of the moon called the Southern Highlands, near the ones named after Christopher Clavius, the first great Jesuit mathematician, and Christopher Scheiner, who was sent to the court of Vienna in 1633 so that Kircher could stay in Rome.

It's not one of the biggest lunar features, but it is big. The crater floor, a smooth, empty plain, stretches forty-five miles across. Kircher the crater has been called "remarkable" for its "very lofty rampart," and though its walls are "somewhat deformed," on the south side they rise up nearly fifteen thousand feet.

These things . . . have been communicated for the serious reader. Many other things could have been brought forth and many other things have already been well described, which I thought I should not repeat. . . . Goodbye, reader, please excuse any errors.

—*China Illustrated*, 1667

ACKNOWLEDGMENTS

Researching and writing one little book about the author of seven million Latin words has a way of putting things in perspective. Like all (self-absorbed) writers, I sometimes imagined I was alone, but in fact I've relied on so many people.

I have endless admiration for the professional scholarship on Kircher, and I'm indebted to the historians behind it. Some have devoted years to a single aspect of his work. I'm especially grateful to those who have written in English, including Martha Baldwin, Paula Findlen, the late John Fletcher, Joscelyn Godwin, Michael John Gorman, the late P. Conor Reilly, S.J., Ingrid D. Rowland, and Daniel Stolzenberg. Their publications, cited in the following pages, are strongly recommended for more in-depth reading.

This book couldn't have been written without access to the resources of many human-being-run entities, including the Archives of the Pontifical Gregorian University in Rome through the Athanasius Kircher Correspondence Project; the Biblioteca Medicea Laurenziana in Florence; the Biblioteca Nazionale Centrale in Rome; the Bibliothèque Nationale de France; the Early English Books Online Text Creation Partnership; the European Cultural Heritage Online project; Google

Books; the Herzog August Bibliothek in Wolfenbüttel; the Internet Archive; the Kircher project at the University of Lucerne; the Museo Galileo; the New York Public Library and its Science, Industry and Business Library; the Royal Society; and the Stanford University Libraries.

Many thanks are owed to Martha Ambrosino, Kathleen Archer, Andreas Armann, Julia Bauerlein, Katherine Bouton, Sarah Bowlin, Daniel D'Addario, Joshua Foer, Max Glassie, Howard Gray, S.J., Yvonne Hicks, Rob Hoerburger, Susannah Jacob, Dr. Berthold Jäger, Bret Anthony Johnston, Meredith Kaffel, Sophie Lvoff, Gerry Marzorati, Giuseppe Mimmo, Alessandro Orlandi, Tom Reiss and Julie Just, Wilhelm Ritz, Rosario Salamone, Sarah Smith, Roberto Tronchin, Sarah Danziger Valentino, and Kornelia Wagner.

The incomparable Laura Bauerlein sent me research and translations from Rome over a long period. The outstanding Camille Silberman sent me translations from Florence and other places. Carey Smith, chair of the classics department at Georgetown Prep, my Jesuit high school, took on the translation of Kircher's autobiography, and kept doing a brilliant job on everything I threw at him. Rachel Nolan provided the finest kind of editorial expertise. Michael John Gorman and Alex Star were kind enough to give me their very valuable comments on the manuscript.

Any errors are mine, but because he asked me to write about Kircher in the first place—for an annual called *The Ganzfeld*, now, sadly, defunct—Dan Nadel is to blame for the entire thing. Violaine Huisman was somehow able to see this book from the start, and can take any credit. Charlotte Sheedy got me through it with uncommon wisdom and generosity. Jake Morrissey guided me with exceptional intelligence and wit. Geoff Kloske bestowed enthusiasm, patience, and more patience on the project. Alexandra Cardia, Anna Jardine, and others at Riverhead have done a truly great job. Thank you.

I'm very grateful to family and friends for their tolerance, love, and support over the years, especially Marian Brown, Claire and Gene Carlin,

Liz and Paul Doucette, Max and Renée Drake, Jeff Glassie and Julie Littell, Tom Glassie, Nancy Green and Steve Saraisky, Scott Hensley, Laurence Master, Yalda Nikoomanesh, Elise Pettus, Ted Pewett, John Taft, and my late great brother Don. My daughter, Natalie, made the sun come out.

Deepest appreciation is reserved for my mother, Claire Buhr Glassie Scrivener, who has done so much, and for the memory of my father, Donelson Caffery Glassie, who once actually stood on the top of the Washington Monument (a great obelisk), and in whose workshop (or cubiculum) I learned so many fascinating things.

NOTES

Apologetic Forerunner to This Kircherian Study

Page

xiii Kircher's account of his early life: Athanasius Kircher, *Vita Admodum Reverendi P. Athanasii Kircheri, Societ. Jesu: Viri Toto Orbe Celebratissimi*, in Hieronymus Ambrosius Langenmantel, *Fasciculus Epistolarum Adm. RP Athanasii Kircheri* (Augsburg: Utzschneider, 1684), hereafter *Vita*. On the likely year of composition, see Nikolaus Seng, trans., *Selbstbiographie des P. Athanasius Kircher aus der Gesellschaft Jesu* (Fulda, Germany: Verlag der Fuldaer Aktiendruckerei, 1901), p. 48.

xiv "It is not the writer's intention": Rene Taylor, "Hermetism and Mystical Architecture in the Society of Jesus," in Rudolf Wittkower and Irma B. Jaffe, eds., *Baroque Art: The Jesuit Contribution* (New York: Fordham University Press, 1972), p. 82; cited in Joscelyn Godwin, *Athanasius Kircher: A Renaissance Man and the Quest for Lost Knowledge* (London: Thames & Hudson, 1979), p. 5.

xvi "Europe's mind was blown": Lawrence Weschler, *Mr. Wilson's Cabinet of Wonder* (New York: Pantheon, 1995), p. 80.

xvii "his works in number, bulk, and uselessness": John Ferguson, *Bibliotheca Chemica: A Catalogue of the Alchemical, Chemical and Pharmaceutical Books in the Collection of the Late James Young of Kelly and Durris* (Glasgow: J. Maclehose and Sons, 1906), vol. 1, p. 468, in José Alfredo

Bach, "Athanasius Kircher and His Method: A Study in the Relations
of the Arts and Sciences in the Seventeenth Century," Ph.D. diss., University of Oklahoma, 1985, pp. 47–48, n. 75.

Chapter 1. Incapable of Resisting the Force

3 Kircher's parents, the circumstances of his birth and youth: *Vita*,
   pp. 1–6.
4 Balthasar von Dernbach and the witch trials: Marc R. Forster, "Review
   of Gerrit Walther, *Abt Balthasars Mission: Politische Mentalitäten, Gegen-
   reformation und eine Adelsverschwörung im Hochstift Fulda*," H-German,
   H-Net Reviews, March 2004, http://www.h-net.org/reviews; Berthold
   Jäger, "Zur Geschichte der Hexenprozesse im Stift Fulda," *Fuldaer
   Geschichtsblätter* 73 (1997), pp. 7–64.
4 The birth of the ninth: In his autobiography, Kircher gives 1602 as the
   year of his birth, but in a few other instances gives it as 1601. See Paula
   Findlen, "The Last Man Who Knew Everything . . . Or Did He?"
   in Paula Findlen, ed., *Athanasius Kircher: The Last Man Who Knew
   Everything* (New York: Routledge, 2004), p. 43, n. 2.
6 "Westerners at this time": Raffaella Sarti, *Europe at Home: Family and
   Material Culture, 1500–1800*, trans. Allan Cameron (New Haven: Yale
   University Press, 2002), p. 113.
6 Description of a typical seventeenth-century home: Ibid., pp. 97, 99–101.
7 "not ordinary aptitude": *Vita*, p. 7.
7 "a wonderful intellect": Andrea Nicoletti, *Della Vita di Papa Urbano
   VIII e Historia del Suo Pontificato*, in Leopold von Ranke, *The History of
   the Popes: Their Church and State, and Especially of Their Conflicts with
   Protestantism in the Sixteenth & Seventeenth Centuries*, trans. E. Foster
   (London: Henry G. Bohn, 1848), vol. 3, Appendix No. 120, p. 405.
7 "thousands of books": *Vita*, p. 2.
7 "entered orders of various religions": Ibid., p. 4.
7 "the world according to its divisions": Ibid., p. 7.
8 Mill wheel story: Ibid., pp. 11–12.
8 Horse race story: Ibid., pp. 13–14.
9 "Very many who want to be counted": In Robert Schwickerath, S.J.,

Jesuit Education: Its History and Principles Viewed in Light of Modern Educational Problems (St. Louis: B. Herder, 1903), p. 147.

10 "propagation of the faith": Ibid., p. 77.

10 Growth of Jesuit schools and seminaries: Allan P. Farrell, S.J., introduction to *The Jesuit Ratio Studiorum of 1599* (Washington, D.C.: Conference of Major Superiors of Jesuits, 1970), p. iii; see also Jonathan Wright, *God's Soldiers: Adventure, Politics, Intrigue, and Power—A History of the Jesuits* (New York: Doubleday, 2004), p. 60.

10 Ignatius of Loyola and humanism: George W. Traub, S.J., *An Ignatian Spirituality Reader* (Chicago: Loyola Press, 2008), pp. 11–14.

10 "cosmopolitan, nonconformist, elitist": Forster, "Review of Gerrit Walther."

10 "anoint their pupils": In Schwickerath, *Jesuit Education*, pp. 147–148.

11 "concerned himself with this one thing": *Vita*, p. 9.

11 "spurn all those things": Ibid., p. 10.

11 "I had heard that a tragedy was being staged": Ibid., p. 16.

12 "are racked and tortured": John Taylor, *Taylor His Travels: From the Citty of London in England, to the City of Prague in Bohemia* (London: Nicholas Okes, 1620), p. [B4] verso.

12 Spessart Forest story: *Vita*, pp. 17–18.

13 "a spirit unrelentingly devoted": Ibid., p. 8.

14 "exceptional joy": Ibid., p. 18.

14 Ice-skating story: Ibid., pp. 19–21.

14 Scabies versus chilblains: P. Conor Reilly, S.J., *Athanasius Kircher, S.J.: Master of a Hundred Arts, 1602–1680* (Rome: Edizioni del Mondo, 1974), p. 27.

15 "ardent pleas": *Vita*, p. 18.

15 "stomach trouble or headache trouble": In Wright, *God's Soldiers*, p. 48.

15 "lest the diseases become known to my superiors": *Vita*, p. 21.

16 "carried day and night in waggons" . . . "Gudgeons newly taken": Taylor, *Taylor His Travels*, pp. B–[B4], [C4]–D. When Taylor asked about the excessive salting, he was told that the beer was so bad it could not be consumed "except if their meat were salted extraordinarily."

16 "Only He who knows the hearts of all" . . . "spread about and stir up": *Vita*, pp. 22–24.

Chapter 2. Inevitable Obstacles

18 "gravity and malice" . . . "how those in hell are licked around": Ignatius of Loyola, *Personal Writings*, ed. and trans. Joseph A. Munitiz and Philip Endean (New York: Penguin, 1997), pp. 296–299.

19 "The most practical and safest": Ibid., p. 301.

20 "practiced by the evil leader": Ibid., p. 311.

20 "under the appearance of good": Ibid., p. 285.

20 "disordered attachments": Ibid., p. 283.

20 "I have it if I find myself at a point": Ibid., p. 315.

21 "I did not dare to reveal my talent": *Vita*, p. 27.

21 "This silence and masking of my ability": Ibid., pp. 28–29.

22 Aristotle: Aristotle, *Physics*, trans. R. P. Hardie and R. K. Gaye ([1930] The Internet Classics Archive, http://classics.mit.edu/Aristotle/physics.html); Michael Fowler, "Aristotle," lecture, University of Virginia, September 3, 2008, http://Galileo.phys.Virginia.EDU/classes/109N/lectures/aristot2.html; Tom Sorell, *Descartes: A Very Short Introduction* (Oxford: Oxford University Press, 2000), pp. 40–41.

23 "shall not depart from Aristotle": *The Jesuit Ratio Studiorum of 1599*, trans. Allan P. Farrell, S.J. (Washington, D.C.: Conference of Major Superiors of Jesuits, 1970), p. 40.

23 "It be a matter of daily observation": Francesco Redi, *Experiments on the Generation of Insects* (1668), trans. Mab Bigelow (Chicago: Open Court, 1909), p. 27.

24 "without any union of parents": In Harry Beal Torrey, "Athanasius Kircher and the Progress of Medicine," *Osiris* 5 (1938), p. 263.

24 Public debate on Aristotle: Sorell, *Descartes*, p. 26.

24 "a new crisis arose": *Vita*, p. 29.

25 "Gottes Freund, der Pfaffen Feind": Ludwig Pastor, *The History of the Popes from the Close of the Middle Ages: Drawn from the Secret Archives of the Vatican and Other Original Sources*, trans. Frederick Ignatius Antrobus, Ralph Francis Kerr, and Ernest Graf (London: Kegan Paul, Trench, Trubner, 1938), vol. 27, p. 240.

25 "He possessed little qualification" . . . "the most famous of them": C. V. Wedgwood, *The Thirty Years War* (New York: New York Review of Books Classics, 2005 [1938]), pp. 145–146.

25 "He issued startling letters": Ibid., pp. 147–148.

25 "lest there be a violent attack": For Kircher's telling of the story, see *Vita*, p. 30.

25 Soon a crowd of Paderborn's Protestants: Reilly, *Athanasius Kircher, S.J.*, pp. 29–30.

26 "And since the enemy was beginning": *Vita*, p. 30.

26 "The winter at that time": On Kircher's escape and journey, see ibid., pp. 31–34, 38, 39.

26 "took subjects prisoner": "Letter of Archbishop Ferdinand of Cologne (July 6, 1622)," in Tryntje Helfferich, ed., *The Thirty Years War: A Documentary History* (Indianapolis: Hackett, 2009), pp. 61–62.

26 "young Dukes of Brunswick": Wedgwood, *Thirty Years War*, p. 151.

26 "wandered in the most dense forest and fields": *Vita*, p. 31.

28 *Washington Crossing the Delaware*: Dr. Bernard J. Cigrand, "Washington Crossing Rhine, Not Delaware; Leutze's Famous Painting Really Represents the German River, and German Soldiers Were Used as Models—American Pupil Aided Artist to Get Proper Uniforms," *The New York Times Magazine*, February 17, 1918, p. 69.

29 "Two altogether inevitable obstacles": *Vita*, pp. 41–43.

30 "stiffened by the vehemence": *Vita*, p. 44.

Chapter 3. A Source of Great Fear

31 "received and restored": *Vita*, p. 45.

32 "superficial quantitative properties": Peter Dear, *Revolutionizing the Sciences: European Knowledge and Its Ambitions, 1500–1700* (Princeton, N.J.: Princeton University Press, 2001), p. 66.

33 "most rapidly and easily destroyed": In Michael J. Gorman, *The Scientific Counter-Revolution: Mathematics, Natural Philosophy and Experimentalism in Jesuit Culture 1580–c1670* (Florence: European University Institute, 1998), p. 34.

34 "outstandingly erudite": Ibid., p. 36.

34 "distributed in various nations and kingdoms": Ibid.

34 "nearly one thousand times larger": Galileo Galilei, *Discoveries and*

Opinions of Galileo, trans. Stillman Drake (Garden City, N.Y.: Double-day, 1957), p. 29.

34　"the moon is not robed in a smooth and polished surface": Ibid., p. 28.

35　"troublesome to operate": In Gorman, *The Scientific Counter-Revolution*, p. 66.

35　"not sufficiently certain": Ibid., p. 68.

37　"The time came": *Vita*, p. 46.

38　"See! How the shadow flies": In Bach, "Athanasius Kircher and His Method," p. 43, n. 43; on Kircher's sundials, see also John Fletcher, "Astronomy in the Life and Correspondence of Athanasius Kircher," *Isis* 61 (1970), p. 53.

39　"huge illusion of their vast antiquity": Frances A. Yates, *Giordano Bruno and the Hermetic Tradition* (London: Routledge & Kegan Paul, 1964), p. 21.

39　Ficino's translations of Hermes in twenty editions: Erik Iversen, *The Myth of Egypt and Its Hieroglyphs in European Tradition* (Princeton, N.J.: Princeton University Press, 1993 [1961]), p. 61.

39　"In the middle of all these things sits the sun": The Latin is in Yates, *Giordano Bruno*, p. 154.

39　"long before the sages": Ibid., p. 11.

40　"the first author of theology": Ibid., p. 14.

40　"foresaw the ruin of the antique religion": In Brian A. Curran, "The Renaissance Afterlife of Ancient Egypt (1400–1650)," in Tim Champion and John Tait, eds., *Encounters with Ancient Egypt: The Wisdom of Egypt: Changing Visions Through the Ages* (London: UCL Press, 2003), p. 109.

40　"as though he were himself a god": In Yates, *Giordano Bruno*, p. 35.

41　"If rational": Giovanni Pico della Mirandola, "Oration on the Dignity of Man," trans. E. L. Forbes, in Ernst Cassirer, Paul O. Kristeller, and John H. Randall, eds., *The Renaissance Philosophy of Man: Selections in Translation* (Chicago: University of Chicago Press, 1948), p. 225.

41　"[Seventy-two] Cabalistic Conclusions": The English translation of Pico della Mirandola is from S. A. Farmer, *Syncretism in the West: Pico's 900 Theses (1486): The Evolution of Traditional Religious and Philosophical Systems* (Tempe, Ariz.: MRTS, 1998), p. 517.

42 "believe that nothing is impossible": In Yates, *Giordano Bruno*, p. 32.

42 "Scarcely any mortal": In Michael John Gorman, "Between the Demonic and the Miraculous: Athanasius Kircher and the Baroque
Culture of Machines," unabridged essay published in abridged form
in Daniel Stolzenberg, ed., *The Great Art of Knowing: The Baroque Encyclopedia of Athanasius Kircher* (Stanford, Calif.: Stanford University Libraries, 2001), http://hotgates.stanford.edu/Eyes/machines/
index.htm.

Chapter 4. Scenic Proceedings

43 "the rudiments of grammar": *Vita*, p. 46.

43 "later he tended to over-compensate": Reilly, *Athanasius Kircher,
S.J.*, p. 33.

44 "I was advised by many to change my religious garb": *Vita*, p. 48.

44 "a certain dark and bristling valley": For the story of Kircher's abduc-
tion by the "heretic horseman," see ibid., pp. 48–52.

45 "the utmost zeal": Ibid., p. 52.

45 "more than 2,000 secrets of medicine": In Paula Findlen, *Possessing
Nature: Museums, Collecting, and Scientific Culture in Early Modern Italy*
(Berkeley: University of California Press, 1994), p. 222.

45 "The Wolf is afraid of the Urchin": Giambattista della Porta, *Natural
Magick, by John Baptista Porta, a Neopolitane: In Twenty Books . . .
Wherein are set forth all the Riches and Delights of the Natural Sciences*
(London: Thomas Young and Samuel Speed, 1658), p. 9.

46 "a Man out of his senses for a day": Ibid., p. 219.

46 "to Adorn Women, and Make them Beautiful": Ibid., p. 233.

47 "A magnificence not to be scoffed at": *Vita*, p. 52.

47 "optical illusions on a grand scale as well as a pyrotechnic display":
Reilly, *Athanasius Kircher, S.J.*, pp. 35, 111.

47 an illuminated flying dragon: Johann Stephan Kestler, *Physiologia
Kircheriana Experimentalis: Qua Summa Argumentorum Multitudine &
Varietate . . .* (Amsterdam: Jansson-Waesberg, 1680), pp. 246–247.

47 "I was exhibiting things" . . . "Several accused me falsely": *Vita*, p. 52.

48 "bad angels": Gorman, "Between the Demonic and the Miraculous."

48 "In order to free myself from this lowly charge": *Vita*, p. 52.

48 "those men departed completely satisfied in every way": Ibid., p. 54.

Chapter 5. Chief Inciter of Action

49 The Prince-Elector of Mainz: It was one of his predecessors who played an important role in the circumstances—some might say caused the circumstances—that started the Reformation. After paying Rome for the privilege of overseeing the sale of indulgences within the German provinces, the Prince-Elector brought on an overeager preacher-salesman, Johann Tetzel, whose range of offerings eventually and infamously included forgiveness for sins not yet committed. That led Martin Luther to nail his ninety-five theses to the door of the church in Wittenberg.

49 "private recreation": *Vita*, pp. 55–56.

50 "When a strong magnet is placed in Peter's breast": Translation in Gorman, "Between the Demonic and the Miraculous."

50 "restores husbands to wives": In Gerrit Verschuur, *Hidden Attraction: The History and Mystery of Magnetism* (New York: Oxford University Press, 1993), p. 8.

50 Medical uses of the lodestone: Martha Baldwin, "Athanasius Kircher and the Magnetic Philosophy," Ph.D. diss., University of Chicago, 1987, pp. 359–363.

51 Gout and the afflicted person's toenails: Ibid., p. 373.

51 "anointed with garlic": William Gilbert, *On the Magnet: Magnetick Bodies Also, and on the Great Magnet of the Earth: A New Physiology, Demonstrated by Many Arguments & Experiments*, trans. S. P. Thompson and the Gilbert Club (London: Chiswick Press, 1900), p. 2.

51 "an experiment with seventy excellent diamonds": Ibid., p. 143.

51 "stupendous implanted vigour": Ibid., p. 68.

51 "the chief inciter of action in nature": William Gilbert, *De Magnete*, trans. P. Fleury Mottelay (New York, 1958), p. 333, in J. A. Bennett, "Cosmology and the Magnetical Philosophy 1640–1680," *Journal for the History of Astronomy* 12 (1981), p. 165.

52 "built all Astronomy": In Bennett, "Cosmology and the Magnetical Philosophy," p. 165.

52 "length, breadth, heights, depths, areas": J. F. Reimann, *Versuch einer Einleitung in die Historiam Literariam . . .* (1708), vol. 4, p. 179, in Fletcher, "Astronomy," p. 54.

52 "delighted to such a marvelous degree": *Vita*, pp. 54–57.

53 twelve major and thirty-eight minor sunspots: Fletcher, "Astronomy," p. 54.

53 "not without wonderment": Athanasius Kircher, *Ars Magna Lucis et Umbrae, in Decem Libros Digesta . . .* (Rome: Scheus, 1646), pp. 6–8, in Bach, "Athanasius Kircher and His Method," p. 43, n. 49.

53 "I was utterly occupied": *Vita*, p. 57.

54 The plague in Prague: Wedgwood, *Thirty Years War*, p. 190.

54 insides eaten away by a huge worm: Ibid., p. 202.

54 "Arabia, Palestine, Constantinople": The Latin is in Daniel Stolzenberg, "Egyptian Oedipus: Antiquarianism, Oriental Studies, and Occult Philosophy in the Work of Athanasius Kircher," Ph.D. diss., Stanford University, 2004, p. 78, n. 20.

54 "For the love of God": Ibid., p. 78, n. 21; translation amended.

55 "Was it by chance": *Vita*, p. 58.

55 "Instantly carried away with curiosity": Ibid., pp. 58–59.

56 "It was the opinion of the ancient theologians": In Curran, "Renaissance Afterlife of Ancient Egypt," p. 109.

56 "From that very moment": *Vita*, pp. 59–60.

57 "high peace resided over the Catholics": Ibid., p. 63.

57 "wretched plight": Friedrich Spee (1631) in Alan C. Kors and Edward Peters, eds., *Witchcraft in Europe, 400–1700: A Documentary History* (Philadelphia: University of Pennsylvania Press, 2001), p. 357.

58 "highly derivative": Baldwin, "Athanasius Kircher and the Magnetic Philosophy," p. 140.

58 "a primary and radical vigor": On Kircher's first book, see ibid., pp. 140–141.

58 "new and sudden whirlwinds of wars": *Vita*, p. 61.

58 The march of Gustavus Adolphus: Wedgwood, *Thirty Years War*, pp. 295–296.

58 "the entire College was dissolved": *Vita*, p. 66.

59 "At Bamberg the bodies lay unburied": Wedgwood, *Thirty Years War*, pp. 320–321.

Chapter 6. Beautiful Reports

60 "Since all things in Germany": *Vita*, p. 67.

61 "an Egyptian sky": In Fletcher, "Astronomy," p. 54.

61 "talent" was "good": The Latin and an English translation is in Stolzenberg, "Egyptian Oedipus," p. 80, n. 26.

61 "strange combination": Reilly, *Athanasius Kircher, S.J.*, p. 40.

62 "fell in with": *Vita*, p. 68.

62 On Peiresc: See Peter N. Miller, *Peiresc's Europe: Learning and Virtue in the Seventeenth Century* (New Haven: Yale University Press, 2000). See also Pierre Gassendi, *The Mirrour of True Nobility & Gentility, Being the Life of the Renowned Nicolaus Claudius Fabricius, Lord of Peiresk*, trans. William Rand (London: J. Streater, for Humphrey Moseley, 1657).

63 "On those dayes on which the Posts": Gassendi, *Mirrour of True Nobility*, pp. 162–163.

63 "could not have been very intimate": Stolzenberg, "Egyptian Oedipus," p. 23. See Peiresc to Dupuy, October 11, 1632, Philippe Tamizey de Larroque, ed., *Lettres de Peiresc* (Paris: Imprimerie Nationale, 1888–1898), vol. 2, p. 359; and Peiresc to Gassendi, April 5, 1633, Tamizey de Larroque, *Lettres de Peiresc*, vol. 4, pp. 300–301. Peiresc was not the first person, nor was he the last, to have trouble getting his name straight. The problem was apparently exacerbated by the existence of another Jesuit, named Balthazar Kitzner. See Stolzenberg, "Egyptian Oedipus," p. 24, n. 4.

64 "Rabbi Barachias Nephi of Babylon": Peiresc to Gassendi, Aix, April 5, 1633, Tamizey de Larroque, *Lettres de Peiresc*, vol. 4, pp. 300–301.

64 "all written with Hieroglyphick Letters": Gassendi, *Mirrour of True Nobility*, p. 207.

64 "made me more hopeful": Peiresc to Gassendi, March 2, 1633, Tamizey de Larroque, *Lettres de Peiresc*, vol. 4, p. 295; compare Stolzenberg, "Egyptian Oedipus," p. 26.

64 "rare courtesie and affability": Gassendi, *Mirrour of True Nobility*, p. 160.

65 "great pleasure in taking him around": Peiresc to Dupuy, Aix, May 21, 1633, Tamizey de Larroque, *Lettres de Peiresc*, vol. 2, p. 528.

65 "by reason of mice": Gassendi, *Mirrour of True Nobility*, p. 166.

65 "waited upon him": Ibid., p. 164.

65 the bones of a giant: Miller, *Peiresc's Europe*, p. 30.

65 "an evil magician, a doctor": In William Huffman, introduction to William Huffman, ed., *Robert Fludd* (Berkeley: North Atlantic Books, 2001), p. 31.

66 "He has beautiful reports": Peiresc to Dupuy, Aix, May 21, 1633, Tamizey de Larroque, *Lettres de Peiresc*, vol. 2, p. 528.

67 "If the experiment that you describe": Descartes to Mersenne, July 22, 1633, in Paul Tannery and Cornelis de Waard, eds., *Correspondance du P. Marin Mersenne, religieux minime* (Paris: G. Beauchesne, 1932–1988), vol. 3, pp. 457–460.

67 "No such author or text": See Stolzenberg, "Egyptian Oedipus," p. 47.

67 "I am called to Vienna": *Vita*, p. 70. It's not clear whether Kircher had really been designated imperial mathematician in Vienna, or if the emperor just wanted to add another mathematician to his staff. Kircher curiously mentioned in one letter that he had been ordered to Trieste, but he referred to Vienna in others. See Stolzenberg, "Egyptian Oedipus," p. 30, n. 25.

68 "Upon learning this": *Vita*, p. 70.

68 "will surely be delayed": In Stolzenberg, "Egyptian Oedipus," p. 31.

69 Waterwheel story: See *Vita*, p. 73.

70 "one third of an hour after two" . . . "to be a miracle of any kind": Peiresc's account of the demonstration is translated in Thomas L. Hankins and Robert J. Silverman, *Instruments and the Imagination* (Princeton, N.J.: Princeton University Press, 1995), pp. 23–27.

71 "The bother that he made": "Note de Peiresc Après la Visite du P. Kircher," September 3, 1633, *Lettres à Claude Saumaise et à Son Entourage*, ed. Agnes Bresson (Florence: Olschki, 1992), p. 380, in Stolzenberg, "Egyptian Oedipus," p. 35.

71 "I discovered it unfitting": Ibid., p. 381. The French is in Stolzenberg, "Egyptian Oedipus," p. 36, n. 39.

72 "he refused to admit it": Ibid., pp. 381–382, in Stolzenberg, "Egyptian Oedipus," p. 37. Translation amended.

72 "was suffused with joy": *Vita*, p. 69.

72 Kircher's account of his journey to Vienna: Ibid., pp. 74–86.

74 "presence alone seems to have": Godwin, *Athanasius Kircher*, p. 13.

Chapter 7. Secret Exotic Matters

77 Kircher's travel plans and reassignment to Rome: See Stolzenberg, "Egyptian Oedipus," p. 34, n. 32; p. 38, n. 45.

78 magnificence and filth frequently competed: Joseph Forsyth, *Remarks on Antiquities, Arts, and Letters: During an Excursion in Italy, in the Years 1802 and 1803* (London: J. Murray, 1816), p. 167.

79 "rare music" was "sung": John Evelyn, *Diary and Correspondence of John Evelyn, F.R.S.: to which is subjoined the private correspondence between King Charles I and Sir Edward Nicholas, and between Sir Edward Hyde, afterwards Earl of Clarendon, and Sir Richard Browne*, ed. William Bray (London: Henry Colburn, 1850), vol. 1, pp. 108–109.

79 Bernini's special effects: Filippo Baldinucci, *The Life of Bernini* (1682), trans. Catherine Engass (University Park: Pennsylvania State University Press, 2006), pp. 83–84.

79 "In Rome, one sees only beggars": In Richard E. Spear, "Scrambling for *Scudi*: Notes on Painters' Earnings in Early Baroque Rome," *The Art Bulletin* 85, no. 2 (June 2003), p. 312.

79 "Being environed with walls": Evelyn, *Diary and Correspondence*, vol. 1, pp. 136–137.

80 "horses, all kinds of corn": Fioravante Martinelli and Henry Cogan, *A Direction for Such As Shall Travell Unto Rome: How They May with Most Ease and Conveniency View All Those Rarities, Curiosities, and Antiquities Which Are to Be Seene There* (London: Henry Herringman, 1654), p. 217.

80 Collegio Romano gave "place to few": Evelyn, *Diary and Correspondence*, vol. 1, p. 132.

80 "I paid my respects" . . . "requested that I tell him": Kircher to Peiresc,

December 1, 1633; the Latin is in Stolzenberg, "Egyptian Oedipus,"
p. 73, n. 4.

82 "to the Vatican library": Ibid., p. 75, n. 6.

83 "News is that there is a Jesuit": Raffaello Maggiotti to Galileo,
March 18, 1634, in *Le Opere di Galileo Galilei* (Florence: G. Barbera,
1968), vol. 16, p. 66.

83 "did not deliberately deceive": Findlen, "The Last Man Who Knew
Everything . . . Or Did He?," p. 19.

83 God "sets limits": *Vita*, p. 91.

84 "lest I fall in with the label of fraud": Ibid., p. 92.

84 "I am hurt": In Stolzenberg, "Egyptian Oedipus," pp. 124–125.

85 multivolume work he planned to call *Universal History*: A complete
transcript and translation of Kircher's outline for this book is in
Stolzenberg, "Egyptian Oedipus," Appendix 2, pp. 343–357.

85 "Father Atanase Kircher is having his Egyptian": Jean Baptiste-Doni to
Mersenne, September 30, 1635, in Paul Tannery and Cornelis de
Waard, eds., *Correspondance du P. Marin Mersenne*, vol. 5 (Paris: Édi-
tions du Centre National de la Recherche Scientifique, 1959), p. 412.

86 It quoted Barachias Nephi: Stolzenberg details the way in which "Barachias Nephi" morphed into the Arab "Barachias Albenephius" or
"Barachias Abenephius" in this and subsequent books. See Stolzenberg, "Egyptian Oedipus," pp. 27, n. 13; 47–48; 51, n. 79. Here, for the
general-interest reader, he will continue to go by "Barachias Nephi."

86 Mount Horeb inscription: See Stolzenberg, "Egyptian Oedipus,"
pp. 118–119.

86 "many arguments": In Ingrid D. Rowland, *The Ecstatic Journey: Athanasius Kircher in Baroque Rome* (Chicago: University of Chicago Press,
2000), pp. 87–88.

86 "unabashedly proclaimed": Curran, "Renaissance Afterlife of Ancient
Egypt," p. 127.

Chapter 8. Habitation of Hell

88 "it happened that": *Vita*, p. 87.

89 Fabio Chigi: See Ingrid D. Rowland, "Etruscan Inscriptions from a

1637 Autograph of Fabio Chigi," *American Journal of Archaeology* 93, no. 3 (1989), pp. 423–428; also see Paula Findlen, "Scientific Spectacle in Baroque Rome: Athanasius Kircher and the Roman College Museum," in Mordechai Feingold, ed., *The Jesuits and the Scientific Revolution* (Cambridge, Mass.: MIT Press, 2002), p. 282.

89 Ramon Llull: See Umberto Eco, *The Search for the Perfect Language*, trans. James Fentress (London: Fontana Press, 1997), pp. 53–69.

90 "enthusiastic capacity for fatiguing detail": Bach, "Athanasius Kircher and His Method," p. 26.

90 Description of the Maltese observatory: Ibid., pp. 51–52, nn. 97, 99, 100; and Rowland, *Ecstatic Journey*, p. 10.

91 asking for a reassignment to Egypt or to the Holy Land: Stolzenberg, "Egyptian Oedipus," pp. 77–78, n. 18.

91 "I found such a Theater of Nature": Athanasius Kircher, *The Vulcano's, Or, Burning and Fire-Vomiting Mountains, Famous in the World, with Their Remarkables: Collected for the Most Part Out of Kircher's Subterraneous World, and Exposed to More General View in English* (London: J. Darby, for John Allen and Benjamin Billingsly, 1669), p. 34.

92 "mariners are wont to allure it": In Reilly, *Athanasius Kircher, S.J.*, p. 67.

92 Kircher's account of the earthquakes in Calabria: *Vita*, pp. 133–147.

92 paid homage to Lucretius, Virgil, Lucan, and Dante: See Findlen, *Possessing Nature*, pp. 184–192.

94 "miracles of subterraneous nature": Kircher, *The Vulcano's*, p. 34; for his account of exploring Vesuvius, see pp. 35–36.

Chapter 9. The Magnet

97 "delayed" him in Rome . . . very dark "states of spirit": In Stolzenberg, "Egyptian Oedipus," pp. 129–130, nn. 193, 195.

98 Instruments in Kircher's cubiculum: Gorman, "Between the Demonic and the Miraculous."

98 limestone stalactites, ostrich eggs . . . and other things: Ingrid D. Rowland, "Athanasius Kircher and the Musaeum Kircherianum," *Humanist Art Review* (n.d.), www.humanistart.net/kircher_idr/kircher.htm.

98 As long as circumstances "held me in Rome": *Vita*, p. 94.

99 "someone in each college of the entire Society": In Baldwin, "Athanasius Kircher and the Magnetic Philosophy," p. 85.

99 one Jesuit in Lithuania: Michael John Gorman, "The Angel and the Compass: Athanasius Kircher's Geographical Project," in Paula Findlen, ed., *Athanasius Kircher: The Last Man Who Knew Everything* (New York: Routledge, 2004), p. 247.

100 "rattle my adversaries' distrust of my work": *Vita*, p. 94.

100 Holy Roman Emperor as new patron: See Stolzenberg, "Egyptian Oedipus," pp. 129–131, nn. 193–196.

101 "We must always maintain that the white I see": Ignatius of Loyola, *Personal Writings*, p. 358.

101 *"absurda, indigna, et intolerabilis"*: In Martha Baldwin, "Magnetism and the Anti-Copernican Polemic," *Journal for the History of Astronomy* 16 (1985), p. 159.

101 "Woe to all iron implements": Athanasius Kircher, *Magnes, sive De Arte Magnetica Opus Tripartum Quo Praeterquam Quod Universa Magnetis Natura* (Rome: Scheus, 1641), p. 544. This is a rather famous line.

101 "wished to philosophize prudently": In Baldwin, "Magnetism and the Anti-Copernican Polemic," p. 160.

102 "that prodigal of nature": In Martha Baldwin, "Kircher's Magnetic Investigations," in Daniel Stolzenberg, ed., *The Great Art of Knowing: The Baroque Encyclopedia of Athanasius Kircher* (Stanford, Calif.: Stanford University Libraries, 2001), p. 28.

102 "coition or union": Robert Fludd, *Mosaicall Philosophy: Grounded Upon the Essentiall Truth or Eternal Sapience: Written First in Latin, and Afterwards Rendred into English* (London, Printed for H. Moseley, 1659), p. 299.

102 "sucketh and attracteth": Ibid., p. 245.

102 Kircher's studies of heliotropic plants: See Baldwin, "Athanasius Kircher and the Magnetic Philosophy," p. 350.

103 "kind of material": In Hankins and Silverman, *Instruments and the Imagination*, p. 29.

103 "pulls what is similar to its own nature": In Baldwin, "Athanasius Kircher and the Magnetic Philosophy," p. 381.

103 "putrid, contagious and noxious to men": Ibid., p. 403.

104 tarantulas and "tarantellas": See ibid., pp. 429ff.

106 vegetable lamb plant of Tartary: Ibid., pp. 341–343.

107 "earned not insignificant applause": *Vita*, p. 94.

107 "a very large volume on the magnet": In Eugenio Lo Sardo, *Iconismi e Mirabilia da Athanasius Kircher* (Rome: Edizioni dell'Elefante, 1999), pp. 13–14.

107 "I am approaching the point": Descartes to Constantijn Huygens, January 5, 1643, in Theo Verbeek and H. J. M Bos, eds., *The Correspondence of René Descartes 1643* (Utrecht: Zeno Institute for Philosophy, 2003), pp. 15–16.

108 "the Magnet by Kircherus": Huygens to Descartes, January 7, 1643, ibid., pp. 17–18.

108 Descartes's own explanation for magnetic attraction and polarity: Park Benjamin, *History of Electricity (The Intellectual Rise of Electricity) from Antiquity to the Days of Benjamin Franklin* (New York: John Wiley & Sons, 1898), pp. 357–361.

109 "flipping through them": Descartes to Huygens, January 14, 1643, in *Correspondence of René Descartes 1643*, pp. 19–20.

Chapter 10. An Innumerable Multitude of Catoptric Cats

110 Procession of the newly elected pope: Evelyn, *Diary and Correspondence*, vol. 1, pp. 130–131.

114 "Father Kircher . . . showed us many singular courtesies": Ibid., p. 108.

114 "far surpassed the competition": Gorman, "Between the Demonic and the Miraculous."

114 "You will exhibit the most delightful trick": Translation ibid.

114 magic lantern: See W. A. Wagenaar, "The True Inventor of the Magic Lantern: Kircher, Walgenstein or Huygens?" *Janus: Archives Internationales pour l'Histoire de la Médecine et pour la Géographie Médicale* 66 (1979), pp. 194–207.

115 dissected the eyeballs of bulls: Reilly, *Athanasius Kircher, S.J.*, p. 82.

115 "We say 'Magna' on account": Athanasius Kircher, *Ars Magna Lucis et Umbrae* . . . ††2 verso. See Bach, "Athanasius Kircher and His Method," p. 231.

116 "For just as the wise men of the Hebrews": Kircher, *Ars Magna Lucis et Umbrae*, in Bach, "Athanasius Kircher and His Method," p. 267, n. 112.

116 "He has established the Sun": Kircher, *Ars Magna Lucis et Umbrae*, pp. 5–6, in Bach, "Athanasius Kircher and His Method," p. 68, p. 87, n. 28.

117 Kircher's theories about fireflies, chameleons, jellyfish: Reilly, *Athanasius Kircher, S.J.*, pp. 77–80.

117 "thickness of the atmosphere": Ibid., p. 84.

117 "rules which must be followed": Ibid., p. 83.

117 why the sky is blue: Kircher, *Ars Magna Lucis et Umbrae*, p. 70, in Bach, "Athanasius Kircher and His Method," p. 91, n. 41.

117 "mites that suggested hairy bears": In Torrey, "Athanasius Kircher and the Progress of Medicine," p. 253.

118 "so tiny that they are beyond": Athanasius Kircher, *Scrutinium Physico-Medicum Contagiosae Luis, Quae Pestis Dicitur . . .* (Rome: Mascardi, 1658), p. 45, in Rowland, *Ecstatic Journey*, p. 105.

118 image-projection . . . a Jesuit in the court wrote him: Johann Gans to Kircher, February 3, 1645, Archivio della Pontificia Università Gregoriana (APUG), ms. 561, fol. 123r; Johann Gans to Kircher, Vienna, June 24, 1645, APUG 561, fol. 135r; "The Correspondence of Athanasius Kircher: The World of a Seventeenth Century Jesuit" (hereafter Athanasius Kircher Correspondence Project), http://archimede.imss.fi.it/kircher.

118 "an eminent man of optics": John Bargrave, *Pope Alexander the Seventh and the College of Cardinals* (1662), ed. James Craigie Robertson ([London]: Camden Society, 1867), p. 134.

118 "Egyptian wanderer": In Rowland, *Ecstatic Journey*, pp. 19–20.

119 "desire of joining to our work on Optics": Athanasius Kircher, Praefatio ad Lectorem, *Phonurgia Nova, sive Conjugium Mechanico-Physicum Artis & Naturae Paranympha Phonosophia Concinnatum . . .* (Kempton, England: Rudolph Dreherr, 1673), p. [C] verso.

119 "tone architecture": Ibid., p. 111.

120 "Should need arise": Ibid., p. 112.

120 infamous cat piano . . . "captured living cats": Kaspar Schott, *Magia Universalis Naturae et Artis, sive Recondita Naturalium et Artificialium*

Rerum Scientia . . . (Würzburg, 1657–1659), pp. 372–373; cited in Hankins and Silverman, *Instruments and the Imagination*, p. 246, n. 2.

122 "notable abuses and faults": Luigi Zenobi and Athanasius Kircher, *The Perfect Musician*, trans. Bonnie J. Blackburn and Leofranc Holford-Strevens (Kraków: Musica Iagellonica, 1995), p. 67.

122 "such wretched compositions": Ibid., p. 73.

122 "the same twittering": Ibid., p. 75.

122 "The mechanical production of music": In Jim Bumgardner, "Kircher's Mechanical Composer: A Software Implementation," *Proceedings of the 2009 Bridges Banff Conference* (Banff, Canada, 2009).

123 "Father Kircher devoured my book": Mersenne to Boulliaud, January 16, 1645, *Correspondance du P. Marin Mersenne*, vol. 13 (ed. Cornelis de Waard, 1977), p. 320.

124 "a more or less direct influence": Roman Vlad, "Kircher: A Knowledgeable Musicologist," in Eugenio Lo Sardo, *Iconismi e Mirabilia da Athanasius Kircher* (Rome: Edizioni dell'Elefante, 1999), p. 66.

Chapter 11. Four Rivers

125 "grounds for praise of God" . . . "elude the empty machinations," *Vita*, pp. 94–95.

126 Kircher's account of the re-erection of the obelisk in Piazza Navona: *Vita*, pp. 96–101.

127 "I would even venture to say": T. A. Marder, "Borromini e Bernini a Piazza Navona," in Christoph Luitpold Frommel and Elisabeth Sladek, eds., *Francesco Borromini: Atti del Convegno Internazionale* (Rome, 2000), p. 144.

128 "Tatu of the Indies": Baldinucci, *Life of Bernini*, p. 37.

129 "One marvels not a little": Ibid., p. 38.

129 "The whole Earth is not solid": Kircher, *The Vulcano's*, p. 3.

129 The fountain as a reflection of Kircher's geology: Rowland, *Ecstatic Journey*, pp. 15, 90.

130 "the lodestone of heaven": Kircher, *Ars Magna Lucis et Umbrae*, p. 2, in Bach, "Athanasius Kircher and His Method," p. 318, n. 84.

130 Bernini "was forever inventing": Simon Schama, *Landscape and Memory* (New York: Alfred A. Knopf, 1995), p. 299.

130 "it is difficult to trace": Ibid., p. 302.

130 Work on the fountain; grain shortage: Giacinto Gigli, *Diario Romano, 1608–1670* (Rome: Tumminelli, 1957), p. 322.

131 "a terrible thing happened": Ibid., pp. 334–335.

131 "he sent to me most eloquent letters": *Vita*, p. 101.

131 Kircher cited a tremor in his right hand: See Stolzenberg, "Egyptian Oedipus," p. 144, n. 10.

132 "the devastation of my entire fatherland": *Vita*, p. 62.

133 "could shoot a hare": In Georgina Masson, *Queen Christina* (New York: Farrar, Straus & Giroux, 1969), p. 115.

133 Descartes in Stockholm: Veronica Buckley, *Christina, Queen of Sweden: The Restless Life of a European Eccentric* (New York: Fourth Estate, 2004), pp. 107–115; A. C. Grayling, *Descartes: The Life and Times of a Genius* (New York: Walker, 2006), pp. 226–234; Masson, *Queen Christina*, pp. 126–132.

134 Queen Christina's Jesuit visitors: See von Ranke, *History of the Popes*, vol. 2, pp. 262–364, and Paolo Casati, "Paolo Casati ad Alessandro VII, sopra la regina di Suecia," in von Ranke, *History of the Popes*, vol. 3, Appendix No. 131, pp. 430–433.

134 "I hope that we shall henceforth": Queen Christina to Kircher, undated, APUG 556, fol. 173r, f. 174r, Athanasius Kircher Correspondence Project, http://archimede.imss.fi.it/kircher/.

Chapter 12. Egyptian Oedipus

135 "half-consumed by rot": Kaspar Schott, "Benevoli Lectori," *Oedipus Aegyptiacus, Hoc Est Universalis Hieroglyphicae Veterum Doctrinae Temporum Iniuria Abolitae Instavratio . . .* (Rome: Mascardi, 1652–1654 [1655]); the Latin is in Stolzenberg, "Egyptian Oedipus," p. 147.

135 "a vast quantity": Ibid., p. 148.

135 "new heavenly bodies": Ibid., p. 157.

135 one Dutch bookseller alone bought five hundred copies: Ibid., p. 145.

136 "as an oracle": In John Fletcher, "A Brief Survey of the Unpublished Correspondence of Athanasius Kircher, S.J. (1602–1680)," *Manuscripta* 13 (1969), p. 150.

136 "one of the most learned monstrosities": Frank E. Manuel, *The Eighteenth Century Confronts the Gods* (Cambridge, Mass.: Harvard University Press, 1959), p. 190, in Baldwin, "Athanasius Kircher and the Magnetic Philosophy," p. 7, n. 1.

136 Chronology of wisdom: Curran, "Renaissance Afterlife of Ancient Egypt," p. 128.

138 "I am fully persuaded": In Stolzenberg, "Egyptian Oedipus," p. 161.

138 "scattered among the chronicles": *Vita*, pp. 57–60.

138 "mysteries of the Egyptians": Latin in Stolzenberg, "Egyptian Oedipus," p. 149, n. 22.

139 "Kircher's he": James Alban Gibbs, Elogium VIII, in Kircher, *Oedipus Aegyptiacus*, p. ++++2.

139 "He has been exceedingly educated": Schott, "Benevoli Lectori"; Latin is in Stolzenberg, "Egyptian Oedipus," p. 147.

140 "Long sections of the *Oedipus*": Stolzenberg, "Egyptian Oedipus," p. 216.

140 "explain doubtful things clearly": Ibid., p. 227.

140 cited "too respectfully": Ibid., p. 221.

142 Barachias Nephi: Ibid., especially pp. 51–59, 65–66, 195, 263.

142 "full plagiarist mode": Ibid., p. 195.

143 "not so much as writing": Don Cameron Allen, "The Predecessors of Champollion," *Proceedings of the American Philosophers Society* 104 (1960), p. 529.

143 "The beneficent Being": Rev. Richard Burgess, "On the Egyptian Obelisks in Rome and Monoliths as Ornaments of Great Cities," read at the Ordinary General Meeting of the Royal Institute of British Architects, May 31, 1858, *Papers Read at the Royal Institute of British Architects* (London: The Institute, 1858), p. 173.

143 "flight of the imagination and learning run mad": Ibid.

144 "fundamental unity of human culture": Curran, "Renaissance Afterlife of Ancient Egypt," p. 102.

Chapter 13. The Admiration of the Ignorant

145 "curious little men": In Aldagisa Lugli, "Inquiry as Collection: The Athanasius Kircher Museum in Rome," *RES: Anthropology and Aesthetics* 12 (Autumn 1986), pp. 109–124.

145 "It happened that I was compelled": Kircher, *Phonurgia Nova*, p. 113.

146 "tail and bones" of a mermaid: Athanasius Kircher, *Arca Noë in Tres Libros Digesta* . . . (Amsterdam: Jansson-Waesberg, 1675), p. 73, in Joscelyn Godwin, *Athanasius Kircher's Theatre of the World: The Life and Work of the Last Man to Search for Universal Knowledge* (Rochester, Vt.: Inner Traditions, 2009), p. 150.

146 "armillary spheres" . . . "an organ, driven by an automatic drum": Giorgio de Sepibus, *Romanii Collegii Musaeum Celeberrimum Cuius Magnae Antiquariae Rei* . . . (Amsterdam: Jansson-Waesberg, 1678), pp. 2–3, in Gorman, "Between the Demonic and the Miraculous."

147 vomiting machines: Ibid. For more on Kircher and Baroque-era regurgitation, see Anthony Grafton, "Magic and Technology in Early Modern Europe," Dibner Library Lecture, October 15, 2002, Smithsonian Institution Libraries (Washington, D.C.: Smithsonian Institution Libraries, 2005), pp. 9–18.

147 "ghosts in the air": Ibid.

148 Delphic Oracle: Kircher, *Phonurgia Nova*, pp. 161–163.

148 "investigation of the learned": Findlen, "Scientific Spectacle in Baroque Rome," p. 262.

149 Kircher's correspondence: See the Athanasius Kircher Correspondence Project, http://archimede.imss.fi.it/kircher/.

149 "It can hardly be said how many inscriptions": Schott, "Benevoli Lectori," in Stolzenberg, "Egyptian Oedipus," p. 149, n. 22.

149 Astronomical reports sent to Kircher: See Fletcher, "Astronomy," pp. 59–60.

151 "finally exalted to the supreme tip": *Vita*, p. 117.

151 "not enjoy what one would call perfect health": Angelo Correr, *Relazione della Corte Romana* (Leiden: Lorens, 1662), p. 69, in Rowland, *Ecstatic Journey*, p. 56.

151 "had so taken upon him the profession": John Bargrave, *Pope Alexander*

the Seventh and the College of Cardinals (1662), ed. James Craigie Robertson ([London]: Camden Society, 1867), p. 7.

151 "liked his company to be gay in reason": Alexis-François Artaud de Montor, *The Lives and Times of the Popes, Including the Complete Gallery of the Portraits of the Pontiffs Reproduced from "Effigies pontificum Romanorum Dominici Basae"* (New York: Catholic Publication Society of New York, 1911 [1836–1843]), vol. 6, p. 105.

151 "literary meetings": "Vita di Alessandro VII," in von Ranke, *History of the Popes*, vol. 3, Appendix No. 135, p. 438.

151 "wished to have Bernini with him every day": Baldinucci, *Life of Bernini*, p. 42.

152 "abrupt and unanticipated departure": Martha Baldwin, "Reverie in Time of Plague," in Paula Findlen, ed., *Athanasius Kircher: The Last Man Who Knew Everything* (New York: Routledge, 2004), pp. 68–69.

152 "no established domicile": In Gale E. Christianson, "Kepler's Somnium: Science Fiction and the Renaissance Scientist," *Science Fiction Studies*, No. 8, vol. 3, part 1 (March 1976), http://www.depauw.edu/sfs/backissues/8/christianson8art.htm.

153 "You are mistaken, and greatly so": In Ingrid D. Rowland, "Athanasius Kircher, Giordano Bruno, and the *Panspermia* of the Infinite Universe," in Paula Findlen, ed., *Athanasius Kircher: The Last Man Who Knew Everything* (New York: Routledge, 2004), p. 194.

153 "a vessel full of celestial dew" . . . "immense Ocean": Athanasius Kircher, *Itinerarium Exstaticum Quo Mundi Opificium* . . . (Rome: Vitale Mascardi, 1656), pp. 122–123.

153 Mars is "harsh with bulges": Ibid., pp. 185–186.

153 Saturn has "horrendous form": Ibid., p. 231.

154 "The whole mass of this solar globe": In Rowland, "Athanasius Kircher, Giordano Bruno," p. 195.

154 "for the most part inane": In Baldwin, "Athanasius Kircher and the Magnetic Philosophy," p. 38.

154 "the offspring of consummate scholarship": In Fletcher, "Astronomy," p. 58.

154 "easily the Phoenix amongst the learned men": Ibid., p. 52.

154 "To be sure, Kircher on occasion reproves": In Rowland, *Ecstatic Journey*, p. 100.

155 "the way a peasant uses his fields": In Buckley, *Christina, Queen of Sweden*, p. 72.

155 "a political invention": Ibid., p. 162.

155 Queen Christina's entry into Denmark: Ibid., p. 163.

156 "Father Athanasius Kircherus the great Mathematician": Galeazzo Gualdo Priorato, *The History of the Sacred and Royal Majesty of Christina Alessandra, Queen of Swedland*, trans. John Burbury (London: T.W., 1658), pp. 428–431.

156 "She has seen everything": In von Ranke, *History of the Popes*, vol. 2, p. 356.

157 "everything from rotten fruit to dead cats": Torgil Magnuson, *Rome in the Age of Bernini: From the Election of Sixtus V to the Death of Urban VIII* (Atlantic Highlands, N.J.: Humanities Press, 1982), p. 190, in Buckley, *Christina, Queen of Sweden*, p. 189.

Chapter 14. Little Worms

158 people "of the highest quality": R. Goodwin and Richard Burdekin, *An Historical Account of the Plague: And Other Pestilential Distempers Which Have Appear'd in Europe* (York, England: R. Burdekin, 1832 [1743]), p. 34.

158 the pestilence had "slithered" inside: Athanasius Kircher, Prooemium ad Lectorem, *Scrutinium Physico-Medicum Contagiosae Luis, Quae Pestis Dicitur* . . . (Rome: Vitale Mascardi, 1658), p. ++ .

159 "with evil signs": Guenter B. Risse, *Mending Bodies, Saving Souls: A History of Hospitals* (New York: Oxford University Press, 1999), p. 192.

159 every year except *two*: According to the medical historian Jean-Noel Biraben; see the review of William Eamon, *Healers and Modern Healing*, in *Renaissance Quarterly*, June 22, 1999.

159 Kircher's descriptions of the symptoms of the plague: in Ralph H. Major, *Classic Descriptions of Disease: With Biographical Sketches of the Authors* (Springfield, Ill.: Charles C. Thomas, 1932), pp. 82–83.

162 "Here you are overwhelmed": In Risse, *Mending Bodies, Saving Souls*, p. 208. See pp. 190–213 for an overview of the plague and the San Bartolomeo pesthouse.

163 "plague most atrocious" . . . "In this state of affairs": Kircher, Prooemium ad Lectorem, *Scrutinium Physico-Medicum Contagiosae Luis*, pp. +++++2.

164 Kircher's account of contagion: Baldwin, "Athanasius Kircher and the Magnetic Philosophy," pp. 387–390.

164 It is "generally known that worms grow": In Major, *Classic Descriptions of Disease*, p. 7.

165 "Take a piece of meat" . . . "if you cut a snake into little pieces": In Torrey, "Athanasius Kircher and the Progress of Medicine," pp. 257–258.

166 "The putrid blood of those affected" . . . "these worms, propagators of the plague": In Charles-Edward Amory Winslow, *The Conquest of Epidemic Disease: A Chapter in the History of Ideas* (Madison: University of Wisconsin Press, 1980 [1943]), pp. 149–152.

167 "a farrago of nonsensical speculation": C. C. Dobell, *Antony van Leeuwenhoek and His 'Little Animals'* (London, 1932), p. 365, in Torrey, "Athanasius Kircher and the Progress of Medicine," p. 248.

167 "undoubtedly the first": Fielding H. Garrison, *An Introduction to the History of Medicine* (Philadelphia, 1929), p. 252, in Torrey, "Athanasius Kircher and the Progress of Medicine," p. 246.

168 "there are many seeds of things": Lucretius, *De Rerum Natura*, trans. William H. D. Rouse and rev. Martin F. Smith (Cambridge, Mass.: Harvard University Press, 1975), 6. 1093–1102, pp. 575–577.

168 "imperceptible particles": Winslow, *Conquest of Epidemic Disease*, p. 132; see pp. 117–143.

169 "the reputation of things Kircherian": For response to *Examination of the Plague*, see John Fletcher, "Medical Men in the Correspondence of Athanasius Kircher," *Janus: Archives Internationales pour l'Histoire de la Médecine et pour la Géographie Médicale* 56 (1969), pp. 263–265.

169 Toad amulet: Baldwin, "Athanasius Kircher and the Magnetic Philosophy," pp. 389–390.

Chapter 15. Philosophical Transactions

170 "a book-making, knowledge-regurgitating machine": Findlen, "The Last Man Who Knew Everything . . . Or Did He?," p. 2.

171 Crosses had mysteriously begun to appear: Stephen Jay Gould, "Father Athanasius Kircher on the Isthmus of a Middle State," in Paula Findlen, ed., *Athanasius Kircher: The Last Man Who Knew Everything* (New York: Routledge, 2004), pp. 233–236; see also Brian L. Merrill, *Athanasius Kircher (1602–1680), Jesuit Scholar: An Exhibition of His Works in the Harold B. Lee Library Collections at Brigham Young University* (Provo, Utah: Friends of The Brigham Young University Library, 1989), p. 36.

172 set down in *New and Universal Polygraphy*: See Nick Wilding, "'If you have a secret, either keep it, or reveal it': Cryptography and Universal Language," in Daniel Stolzenberg, ed., *The Great Art of Knowing: The Baroque Encyclopedia of Athanasius Kircher* (Stanford, Calif.: Stanford University Libraries, 2001), pp. 93–103.

173 mathematical organ: Gorman, *The Scientific Counter-Revolution*, p. 251.

173 "had taken to traveling the world by raft": Ingrid D. Rowland, *The Scarith of Scornello: A Tale of Renaissance Forgery* (Chicago: University of Chicago Press, 2004), p. 32.

173 Annius of Viterbo: See Anthony Grafton, *Defenders of the Text: The Traditions of Scholarship in an Age of Science, 1450–1800* (Cambridge, Mass.: Harvard University Press, 1991), pp. 76–93. Also see Erik Iversen, *The Myth of Egypt and Its Hieroglyphs in European Tradition* (Princeton, N.J.: Princeton University Press, 1993 [1961]), pp. 62–63.

174 Kircher's discovery of the shrine of Mentorella: *Vita*, pp. 105–114.

176 "gentlemen, free and unconfined": Thomas Sprat, *The History of the Royal Society of London*, 2nd ed. (London, 1702), in Dorothy Stimson, "Amateurs of Science in 17th Century England," *Isis* 31, no. 1 (1939), p. 37.

177 "The Club-men have cantonized": G. H. Turnbull, "Samuel Hartlib's Influence on the Early History of the Royal Society," *Notes and Records of the Royal Society* 10 (1953), p. 113.

177 Borges and Foucault: See Jorge Luis Borges, "The Analytical Language of John Wilkins," in *Other Inquisitions 1937–1952*, trans. Ruth L. C. Simms (Austin: University of Texas Press, 1993), p. 101; Michel Fou-

cault, *The Order of Things: An Archaeology of Human Sciences* (New York: Vintage, 1994 [1966]), p. xv.

178 "artificial, mathematical, and magical curiosities": Evelyn, *Diary and Correspondence*, vol. 1, p. 293.

179 "baiting Puritans, place jobbing": Benjamin, *History of Electricity*, pp. 406, 408.

179 Vacuum experiments: W. E. K. Middleton, *The History of the Barometer* (Baltimore: Johns Hopkins University Press, 1964), pp. 10–18, 56–65; see also the Museo Galileo on Schott, http://catalogue.museogalileo .it/biography/GasparSchott.html.

180 "Father Kircher is my particular friend": Southwell to Boyle, March 30, 1661, in Thomas Birch, ed., *Robert Boyle: The Works of the Honourable Robert Boyle, in Six Volumes, to Which Is Prefixed the Life of the Author* (London: J. and F. Rivington, L. Davis, et al., 1772), vol. 6, p. 299.

180 vegetable phoenix that Kircher had put on display: Gorman, *The Scientific Counter-Revolution*, pp. 224–225.

180 Christopher Wren and magnetism: Bennett, "Cosmology and the Magnetical Philosophy," pp. 170–172.

181 ". . . magnetic beam, that gently warms": John Milton, *Paradise Lost*, Book III; see Benjamin, *History of Electricity*, p. 438.

181 Microscopes: Turnbull, "Samuel Hartlib's Influence," pp. 114, 115.

181 "Edges of Rasors": "An Account of Micrographia, or the Physiological Descriptions of Minute Bodies, Made by Magnifying Glasses," *Philosophical Transactions of the Royal Society*, vol. 1 (1665), p. 28.

Chapter 16. Underground World

184 "before the eyes of the curious reader": Mark Waddell, "The World, As It Might Be: Iconography and Probabilism in the *Mundus Subterraneus* of Athanasius Kircher," *Centaurus* 48 (2006), pp. 3–22.

184 "the author was present with great danger": Athanasius Kircher, Index Argumentorum, *Mundus Subterraneus, in XII Libros Digestus* . . . (Amsterdam: Jansson and Weyerstraet, 1664–1665).

185 "are nothing but the vent-holes": Kircher, *The Vulcano's*, p. 5. For Kirch-

er's explanation of volcanoes, oceans, and the structure of the earth, see "The Explication of the Schemes, out of Kircher" (n.p.) and pp. 1–5.

187 "Ripening" of base metals: See Tara E. Nummedal, "Kircher's Subterranean World and the Dignity of the Geocosm," in Daniel Stolzenberg, ed., *The Great Art of Knowing: The Baroque Encyclopedia of Athanasius Kircher* (Stanford, Calif.: Stanford University Library, 2001), p. 42.

188 "The alchemists describe it" . . . "Can one metal really be transmuted into another?": In Reilly, *Athanasius Kircher, S.J.*, pp. 115–120.

189 "The world's seed" . . . "the seed of Nature": In Hiro Hirai, "Athanasius Kircher's Chymical Interpretation of the Creation and Spontaneous Generation," in Lawrence M. Principe, ed., *Chymists and Chymistry: Studies in the History of Alchemy and Early Modern Chemistry* (New York: Science History Publications, 2007), pp. 81–82.

189 "a certain matter that we rightly call 'chaotic'" . . . "I say that a certain material spiritus": Ibid., p. 79.

190 "something" . . . "not as a form": Ibid., pp. 84–85.

191 "It would take a whole journal": In Godwin, *Athanasius Kircher*, p. 84.

191 Kircher "understood erosion": Ibid., p. 84.

191 "Since monstrous animals of this kind": Kircher, *Mundus Subterraneus*, vol. 2, p. 89.

192 Oldenburg . . . "Catalogue of my best books": Noel Malcolm, "The Library of Henry Oldenburg," *The Electronic British Library Journal*, www.bl.uk/eblj, p. 23.

192 "Let it be experimented . . . whether Nitrous water, mixed with common salt": In Reilly, *Athanasius Kircher, S.J.*, p. 106.

193 "'Tis an ill Omen, me thinks": In Gorman, *The Scientific Counter-Revolution*, p. 229.

193 plant-derived purges, donkey's milk: See Martha Baldwin, "The Snakestone Experiments: An Early Modern Medical Debate," *Isis* 86 (1995), p. 410.

194 "a large pigeon" . . . "a fine Naples veil": Redi, *Experiments on the Generation of Insects*, pp. 30–36.

194 "I don't know whether": Ibid., p. 43.

194 "I risked a second": Ibid., p. 64.

194 "I believe": Ibid., pp. 34–35.

195 "no animal of any kind": Ibid., p. 64.

Chapter 17. Fombom

197 "multitude of Fathers": In Martha Baldwin, "The Snakestone Experiments," p. 400; see Athanasius Kircher, *Magneticum Naturae Regnum, sive, Disceptatio Physiologica de Triplici in Natura Rerum* . . . (Amsterdam: Johann Jansson, 1667), pp. 50–58.

197 "When this stone was placed": Athanasius Kircher, *China Illustrata, with Sacred and Secular Monuments, Various Spectacles of Nature and Art and Other Memorabilia* (1677), trans. Charles D. Van Tuyl (Muskogee, Okla: Indian University Press, Bacone College, 1987), p. 73.

198 Kircher and Father Boym: Ibid., compare pp. 73 and 74.

198 "I think that the immutable force of nature": Kircher, *Magneticum Naturae Regnum,* p. 18, in Baldwin, "Athanasius Kircher and the Magnetic Philosophy," p. 27.

199 "ignited wide publicity": Baldwin, "Snakestone Experiments," p. 398.

199 became "the leading advocate": Ibid., p. 396.

199 Egyptian roots of Chinese culture: See Kircher, *China Illustrata,* pp. 122–127, 217.

200 "cold and frozen northern zone" . . . "not counting the royal ministers": Ibid., pp. 159–161.

201 animals of China: Ibid., pp. 184–186.

201 tea "gradually being introduced" . . . "is also used for relieving hangover": Ibid., p. 175.

201 "was probably the single most important": Van Tuyl, "Translator's Foreward [*sic*]," *China Illustrata,* p. i.

202 "had a global reputation": Paula Findlen, "A Jesuit's Books in the New World: Athanasius Kircher and His American Readers," in Paula Findlen, ed., *Athanasius Kircher: The Last Man Who Knew Everything* (New York: Routledge, 2004), p. 329.

202 "Truly without exaggeration": Ibid., p. 337. On Alejandro Favián, see ibid., pp. 335–343.

203 "Upon learning of the matter": For Kircher's account of the Santa Maria sopra Minerva obelisk, see *Vita*, pp. 119–127.

205 "supreme spirit and archetype infuses its virtue": In Godwin, *Athanasius Kircher*, p. 62.

206 Kircher's dream of becoming pope: Kaspar Schott, *Physica Curiosa, sive Mirabilia Naturae et Artis . . .* (Würzburg: Jobus Hertz, 1667; 2nd ed.), pp. 455–456.

206 "morbidly austere": William S. Heckscher, "Bernini's Elephant and Obelisk," *The Art Bulletin* 29, no. 3 (September 1947), p. 181.

208 Tricks played on Kircher: See Fred Brauen, "Athanasius Kircher (1602–1680)," *Journal of the History of Ideas* 43, no. 1 (January–March 1982), pp. 129–134.

208 H. L. Mencken: Johann Burkhard Mencken, *The Charlatanry of the Learned* (*De Charlataneria Eruditorum*, 1715), trans. Francis E. Litz, with notes by H. L. Mencken (New York: Alfred A. Knopf, 1937), p. 44.

Chapter 18. Everything

209 the Jesuits "are the best Men that Live on the Earth": A.F., *The Travels of an English Gentleman from London to Rome: On Foot. Containing, a Comical Description of What He Met with Remarkable in Every City, Town, and Religious House in His Whole Journey* (London: J. How, 1704), pp. 168–169.

210 "Those letters you have sent to me": Hieronymus Langenmantel, *Fasciculus Epistolarum Adm. R.P. Athanasii Kircheri Soc. Jesu . . .* (Augsburg, 1684), pp. 21–24, in Bach, "Athanasius Kircher and His Method," p. 55, n. 135.

210 "I firstly fitted the Church": *Vita*, pp. 114–117.

211 "A person may wish": Ignatius of Loyola, *Personal Writings*, p. 356.

211 *Ars Magna Sciendi*: See Nick Wilding, "'If you have a secret, either keep it, or reveal it': Cryptography and Universal Language," pp. 93–103; Merrill, *Athanasius Kircher (1602–1680), Jesuit Scholar*, p. 56.

212 "the art of arts," "the workshop of the sciences": Athanasius Kircher, *Ars Magna Sciendi: In XII Libros Digesta, Qua Nova & Universali*

Methodo . . . (Amsterdam: Jansson and Weyerstraet, 1669), p. 1, in Bach, "Athanasius Kircher and His Method," p. 193, n. 155.

213 "do not generate fresh questions": Umberto Eco, *The Search for the Perfect Language*, trans. James Fentress (London: Fontana Press, 1997), p. 63.

215 "Of what Use this Doctrine may be": "An Accompt of Some Books," *Philosophical Transactions of the Royal Society*, vol. 4 (1669), pp. 1086–1097.

215 Leibniz as self-conscious and skinny: Benson Mates, *The Philosophy of Leibniz: Metaphysics and Language* (Oxford: Oxford University Press, 1989), p. 32.

216 "Virtually every major scientific": Findlen, "The Last Man Who Knew Everything . . . Or Did He?," p. 9.

216 Wind-powered mining: See Matthew Stewart, *The Courtier and the Heretic: Leibniz, Spinoza, and the Fate of the Modern World* (New York: W. W. Norton, 2006), pp. 206–207, 226–227, 232–233.

216 "the elegance and harmony of the world": Ibid., p. 79.

217 "GREAT MAN," the "greatest man": Leibniz to Kircher, Mainz, May 16, 1670, APUG 559, fols. 166r–166v.; transcribed in Paul Friedländer, "Athanasius Kircher und Leibniz: Ein Beitrag zur Geschichte der Polyhistorie im XVII Jahrhundert," *Rendiconti della Pontificia Accademia Romana d'Archeologia*, ser. 3, vol. 13 (1937), pp. 229–231.

218 "All the alchymists were in arms": Charles MacKay, *Memoirs of Extraordinary Popular Delusions and the Madness of Crowds* (London: Office of the National Illustrated Library, 1852), pp. 186–187.

219 "to assist them with his gold-making art": In Reilly, *Athanasius Kircher, S.J.*, p. 179.

219 "true pupil of this art": Salomone de Blauenstein, *Interpellatio Brevis ad Philosophos Veritatis tam Amatores quam Scrutatores pro Lapide Philosophorum, Contra Antichymsiticum Mundum Subterraneum P. Athanasii Kircheri Jesuitae* (Biel [Switzerland]: Desiderius Suitzius, 1667), title page.

219 Arnold of Villanova: Lynn Thorndike, *A History of Magic and Experimental Science: The Seventeenth Century* (New York: Columbia University Press, 1958), p. 575.

219 "When he read how I had very clearly revealed": In Reilly, *Athanasius Kircher, S.J.*, p. 178.

Chapter 19. Not As It Was

223 Noah's traits and qualities: See Godwin, *Athanasius Kircher*, p. 44.

223 "not as it was": In Godwin, *Athanasius Kircher's Theatre of the World*, p. 120.

223 "flawed beyond belief": Findlen, "The Last Man Who Knew Everything . . . Or Did He?," p. 6.

224 Redi and the snake stone: See Baldwin, "Athanasius Kircher and the Magnetic Philosophy," pp. 397–399.

224 "The principal point of this letter": In Martha Baldwin, "The Snakestone Experiments," p. 413.

225 "speaking trumpet": See Samuel Morland, *Tuba Stentoro-Phonica, An Instrument of Excellent Use, As well at Sea as at Land, Invented and variously Experimented in the year 1670 and humbly presented to the Kings Most Excellent Majesty Charles II in the Year, 1671* (London: W. Godbid, for M. Pitt, 1671).

226 acoustical tube "might extend itself": For Kircher's experiments with the acoustical tube on Mentorella, see Kircher, *Phonurgia Nova*, pp. 113–115.

229 Relocation of the museum: See R. Garrucci, "Origini e Vicende del Museo Kircheriano dal 1651 al 1773," *La Civiltà Cattolica*, ser. 10, vol. 12 (1879), pp. 727–739; Roberta Rezzi, "Il Kircheriano, da Museo d'Arte e di Meraviglie a Museo Archeologico," in Maristella Casciato, Maria Grazia Ianniello, and Maria Vitale, eds., *Enciclopedismo in Roma Barocca: Athanasius Kircher e il Museo del Collegio Romano tra Wunderkammer e Museo Scientifico* (Venice: Marsilio, 1986), pp. 295–302.

229 "Thoroughly animated" . . . "For should these things": The Latin is in Garrucci, "Origini e Vicende del Museo Kircheriano," pp. 729–731.

230 "Does the conference of learned persons please you?": In Findlen, *Possessing Nature*, p. 132.

230 French king himself had "deep respect" for his work: Fletcher, "A Brief Survey of the Unpublished Correspondence," pp. 155, 157.

230 Letters continued to arrive: Fletcher, "Medical Men in the Correspondence," pp. 270, 273, 274; John Fletcher, "Claude Fabri de Peiresc and the Other French Correspondents of Athanasius Kircher (1602–1680)," *Australian Journal of French Studies* 9 (1972), p. 264.

230 rector of the Jesuit college of Vilnius: Fletcher, "A Brief Survey of the
     Unpublished Correspondence," p. 156.
231 "various practical problems": Findlen, *Possessing Nature*, p. 92. On
     Kircher and the ark, see Olaf Breidbach and Michael T. Ghiselin,
     "Athanasius Kircher (1602–1680) on Noah's Ark: Baroque 'Intelligent
     Design' Theory," *Proceedings of the California Academy of Sciences* 57, no.
     36 (December 28, 2006).
232 "born from rot": In Godwin, *Athanasius Kircher's Theatre of the World*,
     p. 120.
232 "upper part has the sex and appearance": In Reilly, *Athanasius Kircher,
     S.J.*, p. 169.

Chapter 20. Immune and Exempt

233 "by now is old": Baldigiani to Redi, April 1, 1675, Biblioteca Medicea
     Laurenziana (BML), Redi 219, f. 148/100v.
233 "play a grand joke": Baldigiani to Redi, December 16, 1674, BML, Redi
     219, f. 141/97r.
234 "written a long, rather questionable response": Baldigiani to Redi, n.d,
     BML, Redi 219, f. 200/142.
234 "Prof. Kircher is as obstinate as ever": Ibid.
234 "Yesterday morning he had his last communion": Baldigiani to Redi,
     Rome, March or May 10, 1675, BML, Redi 219, f.164/110r.
235 "a solitary and little-known man": Baldigiani to Redi, date unclear, 1677,
     BML, Redi 219, f. 183/126v.
235 "never shown himself": Baldigiani to Redi, n.d., BML, Redi 219,
     f. 200/142.
235 "First they pardoned themselves": Baldigiani to Redi, BML, February
     15, 1677, Redi 219, f. 180/123v.
236 Possibility of feeble hens: On Petrucci, see Baldwin, "The Snakestone
     Experiments," p. 416.
236 "reckless and impudent slanderers" . . . "all Celebrated future centuries":
     Gioseffo Petrucci, *Prodomo Apologetico alli Studi Chircheriani* (Amster-
     dam: Jansson and Waesberg, 1677), pp. 3–4.

237 "constantly insisted on the marvelous virtues": Petrucci, *Prodomo Apologetico*, p. 21.

237 "He did not go according to the sentiments": Ibid., p. 22.

237 "The works of nature are prodigious": In Baldwin, "Athanasius Kircher and the Magnetic Philosophy," p. 401.

237 "I have never understood": Galileo, *Discoveries and Opinions*, p. 231.

238 only two were sold: See Baldigiani to Redi, August 14, 167[7?], BML, Redi 219, f. 203/144r.v.

Chapter 21. Mentorella

241 "These days, because of age": Baldigiani to Vincenzo Viviani, July 18, 1678, in Antonio Favaro, "Miscellanea Galileiana Inedita," *Memorie del Reale Istituto Veneto di Scienze, Lettere ed Arti* 22, p. 837.

241 "Many others": In Findlen, "The Last Man Who Knew Everything . . . Or Did He?," p. 3.

241 "Philosophy is written": Galileo, *Discoveries and Opinions*, p. 237.

242 "A shape in space has given way": David Berlinski, *Infinite Ascent: A Short History of Mathematics* (New York: Modern Library, 2008), p. 40.

242 "expound" on "a part of Euclid": Evelyn, *Diary and Correspondence*, vol. I, p. 132.

242 how to square the circle: See Findlen, "The Last Man Who Knew Everything . . . Or Did He?," p. 27.

242 "the Cabalists, Arabs, Gnostics" . . . "genuine and licit": Athanasius Kircher, *Arithmologia, sive, De Abditis Numerorum Mysteriis . . .* (Rome: Varese, 1665), title page.

243 "Mystic Monad or, if you will" . . . "all creatures breathe numbers": Ibid., pp. 239–241.

243 "a reverberating sonic *boom!*": Berlinski, *Infinite Ascent*, p. 45.

243 "You must know that now": In Reilly, *Athanasius Kircher, S.J.*, p. 179.

244 "Decrepit and old, Professor Kircher": Baldigiani to Redi, n.d, BML, Redi 219, f. 204/145r.

245 It was God who "wished that I expend" . . . "And so, the subject matter": *Vita*, pp. 89–91.

246　"second childhood": Reilly, *Athanasius Kircher, S.J.*, p. 180.

246　Preparation of bodies: Jean-Nicolas Gannal, *History of Embalming, and of Preparations in Anatomy, Pathology, and Natural History: Including an Account of a New Process for Embalming*, trans. Richard Harlan (Philadelphia: J. Dobson, 1840). See also William D. Haglund and Marcella H. Sorg, *Forensic Taphonomy: The Postmortem Fate of Human Remains* (Boca Raton, Fla.: CRC Press, 1997), p. 487; Pascale Trompette and Mélanie Lemonnier, "Funeral Embalming: The Transformation of a Medical Innovation," *Science Studies* 22, no. 2 (2009), pp. 9–30.

248　"The track which leads to it": The Reverend Robert Belaney, "Our Lady of Mentorella," *The Ave Maria* 51, no. 13 (Notre Dame, Ind., September 29, 1900).

249　there was a clerk "kept quite busy": Filippo Buonanni, *Musaeum Kircherianum, sive, Musaeum a P. Athanasio Kirchero in Collegio Romano Societatis Jesu . . .* (Rome: Georgius Plachus, 1709), p. 1.

249　"Father Kircher's Cabinet": Maximilien Misson, *A New Voyage to Italy: With Curious Observations on Several Other Countries . . .* (London: R. Bonwicke, J. Tonson, et al., 1714 [1695]), vol. 2, p. 172.

Chapter 22. Closest of All to the Truth

250　"what the modern world's about": David Foster Wallace, *Everything and More: A Compact History of Infinity* (New York: W. W. Norton, 2003), pp. 107–109.

251　"I hope to show—as it were, by my example": In Casper Hakfoort, "Newton's Optics: The Changing Spectrum of Science," in John Fauvel et al., eds., *Let Newton Be!* (New York: Oxford University Press, 1989), p. 86.

251　"who find out, settle & do": In James Gleick, *Isaac Newton* (New York: Pantheon, 2003), p. 127.

252　upholster much of what he owned in crimson: See ibid., pp. 231–232, n. 10.

252　"given the length of the space continuously": In John Stillwell, *Mathematics and Its History* (New York: Springer-Verlag, 1989), p. 278.

253 "began to think of gravity": In the introduction to John Fauvel et al., eds., *Let Newton Be!* (New York: Oxford University Press, 1989), p. 14.

253 "ground and polished glasses": In Paula Findlen, "The Janus Faces of Science in the Seventeenth Century: Athanasius Kircher and Isaac Newton," in Margaret J. Osler, ed., *Rethinking the Scientific Revolution* (Cambridge, England: Cambridge University Press, 2000), p. 230.

253 correspondence between the colors of the spectrum: Penelope Gouk, "The Harmonic Roots of Newtonian Science," in John Fauvel et al., eds., *Let Newton Be!* (New York: Oxford University Press, 1989), p. 118.

253 Years later, Voltaire heard: Bach, "Athanasius Kircher and His Method," p. 91, n. 47. See also Findlen, "The Janus Faces of Science," p. 227.

254 "the peerless alchemist of Europe": Gleick, *Isaac Newton*, p. 99.

254 "The Fire scarcely going out either Night or Day": In Gleick, *Isaac Newton*, p. 219, n. 6.

255 more than a million words on alchemy: R. S. Westfall, "Newton and Alchemy," in Brian Vickers, ed., *Occult and Scientific Mentalities in the Renaissance* (Cambridge, England: Cambridge University Press, 1984), p. 321.

255 Newton and Hermes Trismegistus: Lawrence M. Principe, "The Alchemies of Robert Boyle and Isaac Newton: Alternate Approaches and Divergent Deployments," in Margaret J. Osler, ed., *Rethinking the Scientific Revolution* (Cambridge, England: Cambridge University Press, 2000), pp. 212–213.

255 "penetrates every solid thing": In Betty Jo Teeter Dobbs, *The Janus Faces of Genius: The Role of Alchemy in Newton's Thought* (Cambridge, England: Cambridge University Press, 1991), p. 274.

255 "active principle": Ibid., p. 209.

255 "vulgar" or "brute" . . . "propensity to associate": Isaac Newton, Two incomplete treatises on the vegetative growth of metals and minerals, 1670–1675, NMAHRB Ms. 1031 B, Dibner Library, Smithsonian Institution, online via The Newton Project, http://www.newtonproject.sussex.ac.uk, at The Chymistry of Isaac Newton, http://webapp1.dlib.indiana.edu/newton/index.jsp.

256 "This & only this": In Dobbs, *Janus Faces of Genius*, p. 24. See Findlen, "The Janus Faces of Science," p. 234.

257 "Whether by any Magnetick or whatother Tye": In Bennett, "Cosmology and the Magnetical Philosophy," p. 172.

257 "concerning the inflection of a direct motion": Ibid., p. 173.

257 "I have not been able to discover the cause": In John Henry, "Newton, Matter, and Magic," in John Fauvel et al., eds., *Let Newton Be!* (New York: Oxford University Press, 1989), p. 141.

258 "For many things lead me": Isaac Newton, *The Principia: Mathematical Principles of Natural Philosophy: A New Translation*, trans. I. B. Cohen and Anne Whitman (Berkeley: University of California Press, 1999), p. 382.

259 "occult quality" . . . "a supernatural thing": In Alexandre Koyré, *From the Closed World to the Infinite Universe* (Baltimore: Johns Hopkins University Press, 1968), pp. 268, 253.

259 "explained mathematically and mechanically": Franklin Perkins, *Leibniz: A Guide for the Perplexed* (London and New York: Continuum, 2007), p. 74.

259 "invisible, intangible" . . . "must be a perpetual *Miracle*": In Koyré, *From the Closed World to the Infinite Universe*, p. 268.

259 "Surely it is no coincidence": Findlen, "The Janus Faces of Science," p. 234.

260 "In my opinion the Egyptian system": Ádám Ferencz Kollár, *Ad Petri Lambecii Commentariorum de Augusta Bibliotheca Caes. Vindobonensi Libros VIII. Supplementorum Liber Primus Supplementorum Posthumus* (Vienna: Johann Thomas Edler von Trattern, 1790), p. 357, in Stolzenberg, "Egyptian Oedipus," p. 1.

Chapter 23. The Strangest Development

261 Kircher "had not even dreamed": Gottfried Wilhelm Leibniz, *Philosophical Papers and Letters*, ed. and trans. Leroy E. Loemker (Dordrecht, Netherlands: Kluwer, 1989 [1956]), p. 352.

261 "He understands nothing": Gottfried Wilhelm Leibniz, *Discourse on the Natural Theology of the Chinese* (1716), in Leibniz, *Writings on China*,

ed. and trans. Daniel J. Cook and Henry Rosemont, Jr. (Chicago: Open Court, 1994), p. 133. See Findlen, *Athanasius Kircher*, p. 6.

261 Leibniz and the *I Ching*: Gottfried Wilhelm Leibniz, *Explication de l'Arithmétique Binaire*, 1703. See Eco, *Search for the Perfect Language*, pp. 284–287; Umberto Eco, *Serendipities: Language and Lunacy*, trans. William Weaver (New York: Harvest, 1999), pp. 69–73.

262 unlock the secret of the Egyptian hieroglyphic system . . . by studying Chinese: Don Cameron Allen, "The Predecessors of Champollion," *Proceedings of the American Philosophers Society* 104 (1960), pp. 533–547.

263 "the obelisks were seen to enshrine": Godwin, *Athanasius Kircher*, p. 6.

264 William Butler Yeats: See Neil Mann, "W. B. Yeats and the Vegetable Phoenix," in Warwick Gould, ed., *Influence and Confluence: Yeats Annual, No. 17* (Basingstoke, England: Palgrave Macmillan, 2007), pp. 3–35.

264 She even quoted Rabbi Barachias Nephi: H. P. Blavatsky, *The Secret Doctrine: The Synthesis of Science, Religion and Philosophy* (New York: Theosophical Publishing Company, 1888), p. 362. Mentioned in Stolzenberg, "Egyptian Oedipus," p. 64.

265 a "monk" who "appeared among the mystics": H. P. Blavatsky, *Isis Unveiled: A Master-Key to the Mysteries of Ancient and Modern Science and Theology* (New York: J. W. Bouton, 1877), vol. 1, pp. 208–209.

266 Animal magnetism, he wrote, acts "at a distance" . . . "communicated, propagated": Franz Anton Mesmer, *Mesmerism: Being the Discovery of Animal Magnetism* (1779), trans. Joseph Bouleur (Sequim, Wash.: Holmes, 2009), pp. 12, 26–27.

266 "a look of dignity" . . . "gently down the spine": MacKay, *Memoirs of Extraordinary Popular Delusions*, vol. 1, p. 279.

267 "Without the stunning progress": Verschuur, *Hidden Attraction*, p. vi.

268 "more energy-like": In Richard Panek, "Out There," *The New York Times Magazine*, March 11, 2007, www.nytimes.com/2007/03/11/magazine/11dark.t.html.

268 "Proof will require a lot more information": John Glassie, "A Conversation with Edward O. Wilson," Salon.com, January 14, 2002, http://www.salon.com/2002/01/14/eowilson_2/.

269 "To believe in evolution": James E. Strick, *Sparks of Life: Darwinism and the Victorian Debates over Spontaneous Generation* (Cambridge, Mass.: Harvard University Press, 2000), p. 2.

269 Duchamp and de Chirico: See Lugli, "Inquiry as Collection," p. 124.

270 "Kircher and others imagine": Edgar Allan Poe, "A Descent into the Maelström," *Graham's Magazine*, no. 18 (May 1841), p. 237, online at the Edgar Allan Poe Society of Baltimore website, www.eapoe.org. Kircher's influence on Poe and others is mentioned in Findlen, "The Last Man Who Knew Everything . . . Or Did He?," p. 42.

270 "learned egoist": Jules Verne, *Journey to the Centre of the Earth*, trans. William Butcher (Oxford, England: Oxford University Press, 1992), p. 4.

271 "a sage?": Umberto Eco, *The Island of the Day Before*, trans. William Weaver (New York: Penguin, 1996), pp. 271–272.

271 "gladdened the World to the extreme": Kircher, *Itinerarium Exstaticum*, pp. 1–2.

273 "remarkable" for its "very lofty rampart": Thomas Gwyn Elger, *The Moon: A Full Description and Map of Its Principal Physical Features* (London: George Philip & Son, 1895).

273 "somewhat deformed": John Wilkinson, *The Moon in Close-up: A Next Generation Astronomer's Guide* (Berlin: Springer, 2010), p. 246.

SELECTED SOURCES

"An Accompt of Some Books." *Philosophical Transactions of the Royal Society*, vol. 4 (1669), pp. 1086–1097.

"Account of Athanasii Kircheri China Illustrata." *Philosophical Transactions of the Royal Society*, vol. 2(1667), pp. 484–488.

"An Account of Micrographia, or the Physiological Descriptions of Minute Bodies, Made by Magnifying Glasses." *Philosophical Transactions of the Royal Society*, vol. 1 (1665), pp. 27–32.

Bach, José Alfredo. "Athanasius Kircher and His Method: A Study in the Relations of the Arts and Sciences in the Seventeenth Century." Ph.D. diss., University of Oklahoma, 1985.

Baldinucci, Filippo. *The Life of Bernini* (1682). Trans. Catherine Engass. University Park: Pennsylvania State University Press, 2006.

Baldwin, Martha. "Athanasius Kircher and the Magnetic Philosophy." Ph.D. diss., University of Chicago, 1987.

_____. "Magnetism and the Anti-Copernican Polemic." *Journal for the History of Astronomy* 16 (1985), pp. 155–174.

_____. "Reverie in Time of Plague." In Paula Findlen, ed., *Athanasius Kircher: The Last Man Who Knew Everything*. New York: Routledge, 2004.

_____. "The Snakestone Experiments: An Early Modern Medical Debate." *Isis* 86 (1995), pp. 394–418.

Brauen, Fred. "Athanasius Kircher (1602–1680)." *Journal of the History of Ideas* 43, no. 1 (1982), pp. 129–134.

Breidbach, Olaf, and Michael T. Ghiselin. "Athanasius Kircher (1602–1680) on Noah's Ark: Baroque 'Intelligent Design' Theory." *Proceedings of the California Academy of Sciences* 57, no. 36 (December 28, 2006).

Buonanni, Filippo. *Musaeum Kircherianum, sive, Musaeum a P. Athanasio Kirchero in Collegio Romano Societatis Jesu* . . . Rome: Georgius Plachus, 1709.

Casciato, Maristella, Maria Grazia Ianniello, and Maria Vitale, eds. *Enciclopedismo in Roma Barocca: Athanasius Kircher e il Museo del Collegio Romano tra Wunderkammer e Museo Scientifico.* Venice: Marsilio, 1986.

Curran, Brian A. "The Renaissance Afterlife of Ancient Egypt (1400–1650)." In Tim Champion and John Tait, eds., *Encounters with Ancient Egypt: The Wisdom of Egypt: Changing Visions Through the Ages.* London: UCL Press, 2003.

Dear, Peter. *Revolutionizing the Sciences: European Knowledge and Its Ambitions, 1500–1700.* Princeton, N.J.: Princeton University Press, 2001.

della Porta, Giambattista. *Natural Magick, by John Baptista Porta, a Neopolitane: In Twenty Books . . . Wherein are set forth all the Riches and Delights of the Natural Sciences.* London: Thomas Young and Samuel Speed, 1658.

Eco, Umberto. *The Search for the Perfect Language.* Trans. James Fentress. London: Fontana Press, 1997.

———. *Serendipities: Language and Lunacy.* Trans. William Weaver. New York: Harvest, 1999.

Evelyn, John. *Diary and Correspondence of John Evelyn, F.R.S.: to which is subjoined the private correspondence between King Charles I and Sir Edward Nicholas, and between Sir Edward Hyde, afterwards Earl of Clarendon, and Sir Richard Browne.* Ed. William Bray. London: Henry Colburn, 1850.

Feingold, Mordechai, ed. *Jesuit Science and the Republic of Letters.* Cambridge, Mass.: MIT Press, 2003.

Findlen, Paula. "The Janus Faces of Science in the Seventeenth Century: Athanasius Kircher and Isaac Newton." In Margaret J. Osler, ed., *Rethinking the Scientific Revolution* (Cambridge, England: Cambridge University Press, 2000), pp. 221–246.

———. "Living in the Shadow of Galileo: Antonio Baldigiani (1647–1711), a Jesuit Scientist in Seventeenth-Century Rome." In Maria P. Donato and Jill Kraye, eds., *Conflicting Duties: Science, Medicine, and Religion in Rome, 1550–1750*. London: Warburg Institute, 2009.

———. *Possessing Nature: Museums, Collecting, and Scientific Culture in Early Modern Italy*. Berkeley: University of California Press, 1994.

———. "Scientific Spectacle in Baroque Rome: Athanasius Kircher and the Roman College Museum," in Mordechai Feingold, ed., *The Jesuits and the Scientific Revolution*. Cambridge, Mass.: MIT Press, 2002.

———, ed. *Athanasius Kircher: The Last Man Who Knew Everything*. New York: Routledge, 2004.

Fletcher, John. "Astronomy in the Life and Correspondence of Athanasius Kircher." *Isis* 61 (1970), pp. 52–67.

———. "A Brief Survey of the Unpublished Correspondence of Athanasius Kircher, S.J. (1602–1680)." *Manuscripta* 13 (1969), pp. 150–160.

———. "Claude Fabri de Peiresc and the Other French Correspondents of Athanasius Kircher (1602–1680)." *Australian Journal of French Studies* 9 (1972), pp. 250–273.

———. "Medical Men in the Correspondence of Athanasius Kircher." *Janus: Archives Internationales pour l'Histoire de la Médecine et pour la Géographie Médicale* 56 (1969), pp. 259–277.

Galilei, Galileo. *Discoveries and Opinions of Galileo*. Trans. Stillman Drake. Garden City, N.Y.: Doubleday, 1957.

Garrison, F. H. "Athanasius Kircher and the Germ Theory of Disease." *Science* 31, no. 805 (June 3, 1910), pp. 857–859.

Gassendi, Pierre. *The Mirrour of True Nobility & Gentility, Being the Life of the Renowned Nicolaus Claudius Fabricius, Lord of Peiresk*. Trans. William Rand. London: J. Streater for Humphrey Moseley, 1657.

Gigli, Giacinto. *Diario Romano, 1608–1670.* Rome: Tumminelli, 1957.

Gilbert, William. *On the Magnet: Magnetick Bodies Also, and on the Great Magnet of the Earth: A New Physiology, Demonstrated by Many Arguments & Experiments.* Trans. S. P. Thompson and the Gilbert Club. London: Chiswick Press, 1900.

Godwin, Joscelyn. *Athanasius Kircher: A Renaissance Man and the Quest for Lost Knowledge.* London: Thames & Hudson, 1979.

_____. *Athanasius Kircher's Theatre of the World: The Life and Work of the Last Man to Search for Universal Knowledge.* Rochester, Vt.: Inner Traditions, 2009.

Gorman, Michael John. "The Angel and the Compass: Athanasius Kircher's Geographical Project." In Paula Findlen, ed., *Athanasius Kircher: The Last Man Who Knew Everything.* New York: Routledge, 2004.

_____. "Between the Demonic and the Miraculous: Athanasius Kircher and the Baroque Culture of Machines." Unabridged essay published in abridged form in *The Great Art of Knowing: The Baroque Encyclopedia of Athanasius Kircher,* ed. Daniel Stolzenberg. Stanford, Calif.: Stanford University Libraries, 2001, pp. 59–70. http://hotgates.stanford.edu/Eyes/machines/index.htm.

_____. *The Scientific Counter-Revolution: Mathematics, Natural Philosophy and Experimentalism in Jesuit Culture 1580–c1670.* Florence: European University Institute, 1998.

Hankins, Thomas L., and Robert J. Silverman. *Instruments and the Imagination.* Princeton, N.J.: Princeton University Press, 1995.

Hooke, Robert. *Micrographia: or, Some Physiological Descriptions of Minute Bodies Made by Magnifying Glasses, with Observations and Inquiries Thereupon.* London: J. Martyn and J. Allestry, 1665.

Ignatius of Loyola. *Personal Writings.* Ed. and trans. Joseph A. Munitiz and Philip Endean. New York: Penguin, 1996.

Iversen, Erik. *The Myth of Egypt and Its Hieroglyphs in European Tradition.* Princeton, N.J.: Princeton University Press, 1993 (1961).

Kircher, Athanasius. *Arithmologia, sive, De Abditis Numerorum Mysteriis . . .* Rome: Varese, 1665.

_____. *Ars Magna Lucis et Umbrae, in Decem Libros Digesta* . . . Rome: Scheus, 1646.

_____. *Ars Magna Sciendi: In XII Libros Digesta, Qua Nova & Universali Methodo* . . . Amsterdam: Jansson and Weyerstraet, 1669.

_____. *China Illustrata, with Sacred and Secular Monuments, Various Spectacles of Nature and Art and Other Memorabilia* (1677). Trans. Charles D. Van Tuyl. Muskogee, Okla.: Indian University Press, Bacone College, 1987.

_____. *Itinerarium Exstaticum Quo Mundi Opificium* . . . Rome: Vitale Mascardi, 1656.

_____. *Magnes, sive de Arte Magnetica Opus Tripartum Quo Praeterquam Quod Universa Magnetis Natura* . . . Rome: Scheus, 1641.

_____. *Magneticum Naturae Regnum, sive, Disceptatio Physiologica de Triplici in Natura Rerum* . . . Amsterdam: Johann Jansson, 1667.

_____. *Mundus Subterraneus, in XII Libros Digestus* . . . Amsterdam: Jansson and Weyerstraet, 1664–1665.

_____. *Oedipus Aegyptiacus, Hoc Est Universalis Hieroglyphicae Veterum Doctrinae Temporum Iniuria Abolitae Instavratio* . . . Rome: Mascardi, 1652–1654 [1655].

_____. *Phonurgia Nova, sive Conjugium Mechanico-Physicum Artis & Naturae Paranympha Phonosophia Concinnatum* . . . Kempton, England: Rudolph Dreherr, 1673.

_____. *Scrutinium Physico-Medicum Contagiosae Luis, Quae Pestis Dicitur* . . . Rome: Vitale Mascardi, 1658.

_____. *Vita Admodum Reverendi P. Athanasii Kircheri, Societ. Jesu, Viri Toto Orbe Celebratissimi.* In Hieronymus Ambrosius Langenmantel, *Fasciculus Epistolarum Adm. RP Athanasii Kircheri.* Augsburg: Utzschneider, 1684.

_____. *The Vulcano's, Or, Burning and Fire-Vomiting Mountains, Famous in the World, with Their Remarkables: Collected for the Most Part Out of Kircher's Subterraneous World, and Exposed to More General View in English.* London: J. Darby, for John Allen and Benjamin Billingsly, 1669.

_____, and Johann Stephan Kestler, *Physiologia Kircheriana Experimentalis. Qua Summa Argumentorum Multitudine & Varietate.* Amsterdam: Jansson-Waesberg, 1680.

Koyré, Alexandre. *From the Closed World to the Infinite Universe.* Baltimore: Johns Hopkins University Press, 1968 (1957).

Lo Sardo, Eugenio. *Iconismi e Mirabilia da Athanasius Kircher.* Rome: Edizioni dell'Elefante, 1999.

Lovejoy, Arthur O. *The Great Chain of Being: A Study of the History of an Idea: The William James Lectures Delivered at Harvard University, 1933.* Cambridge, Mass.: Harvard University Press, 1964.

Major, Ralph H. *Classic Descriptions of Disease: With Biographical Sketches of the Authors.* Springfield, Ill.: Charles C. Thomas, 1932.

Mencken, Johann Burkhard. *The Charlatanry of the Learned (De Charlataneria Eruditorum, 1715).* Trans. Francis E. Litz, with notes by H. L. Mencken. New York: Alfred A. Knopf, 1937.

Merrill, Brian. L. *Athanasius Kircher (1602–1680), Jesuit Scholar: An Exhibition of His Works in the Harold B. Lee Library Collections at Brigham Young University.* Provo, Utah: Friends of The Brigham Young University Library, 1989.

O'Malley, John W., ed. *The First Jesuits.* Toronto: University of Toronto Press, 2000.

Osler, Margaret J., ed. *Rethinking the Scientific Revolution.* New York: Cambridge University Press, 2000.

Petrucci, Gioseffo. *Prodomo Apologetico alli Studi Chircheriani.* Amsterdam: Jansson and Waesberg, 1677.

Priorato, Galeazzo Gualdo. *The History of the Sacred and Royal Majesty of Christina Alessandra, Queen of Swedland.* Trans. John Burbury. London: T.W., 1660.

Ranke, Leopold von. *The History of the Popes: Their Church and State, and Especially of Their Conflicts with Protestantism in the Sixteenth & Seventeenth Centuries,* vol. 3. Trans. E. Foster. London: Henry G. Bohn, 1848.

Redi, Francesco. *Experiments on the Generation of Insects* (1668). Trans. Mab Bigelow. Chicago: Open Court, 1909.

Reilly, P. Conor, S.J. *Athanasius Kircher, S.J.: Master of a Hundred Arts, 1602–1680.* Rome: Edizioni del Mondo, 1974.

Rivosecchi, Valerio. *Esotismo in Roma Barocca: Studi sul Padre Kircher.* Rome: Bulzoni, 1982.

Rowland, Ingrid D. "Athanasius Kircher, Giordano Bruno, and the *Panspermia* of the Infinite Universe." In Paula Findlen, ed., *Athanasius Kircher: The Last Man Who Knew Everything.* New York: Routledge, 2004.

_____. *The Ecstatic Journey: Athanasius Kircher in Baroque Rome.* Chicago: University of Chicago Press, 2000.

Shapin, Steven. *The Scientific Revolution.* Chicago: University of Chicago Press, 1996.

Stolzenberg, Daniel. "Egyptian Oedipus: Antiquarianism, Oriental Studies, and Occult Philosophy in the Work of Athanasius Kircher." Ph.D. diss., Stanford University, 2004.

_____, ed. *The Great Art of Knowing: The Baroque Encyclopedia of Athanasius Kircher.* Stanford, Calif.: Stanford University Libraries, 2001.

Strasser, Gerhard F. "Science and Pseudoscience: Athanasius Kircher's *Mundus Subterraneus* and His *Scrutinium . . . Pestis.*" In G. Scholz Williams and Stephan K. Schindler, eds., *Knowledge, Science, and Literature in Early Modern Germany.* Chapel Hill: University of North Carolina Press, 1996.

Szczesniak, Baleslaw. "Athanasius Kircher's *China Illustrata.*" *Osiris* 10 (1952), pp. 385–411.

Taylor, John. *Taylor His Travels: From the Citty of London in England, to the Citty of Prague in Bohemia.* London: Nicholas Okes, 1620.

Torrey, Harry Beal. "Athanasius Kircher and the Progress of Medicine." *Osiris* 5 (1938), pp. 246–275.

Waddell, Mark. "The World, As It Might Be: Iconography and

Probabilism in the *Mundus Subterraneus* of Athanasius Kircher."
Centaurus 48 (2006), pp. 3–22.

Wedgwood, C. V. *The Thirty Years War*. New York: New York Review of
Books Classics, 2005.

Wright, Jonathan. *God's Soldiers: Adventure, Politics, Intrigue, and Power—
A History of the Jesuits*. New York: Doubleday, 2004.

Yates, Frances A. *Giordano Bruno and the Hermetic Tradition*. London:
Routledge & Kegan Paul, 1964.

INDEX

ILLUSTRATION CREDITS